YOUR LOVING SON, TY

Your Loving Son, Ty

A World War II Story of Hope and Horror in the Pacific

JODY BECK

Jody Beck

ISBN: 979-8-9874142-0-0

To my parents, who raised five children. They told us to aim for our dreams and were mildly surprised when we did.

Contents

Your Loving Son, Ty

Introduction

This story is about my mother's cousin, Madsen Cobb Kokjer, a U.S. Army Air Forces pilot in World War II. Everyone called him Ty. His mother's maiden name was Cobb, and baseball legend Ty Cobb was at the height of his fame when Ty was born, in 1919, in Kearney, Nebraska. The nickname stayed with him for the rest of his brief life.

Ty was a friendly, outgoing boy and was close to his parents. When it was time for him to go to college, he did not go far, choosing the University of Nebraska in Lincoln. But after five semesters, he was bored. Eager for some adventure, he left after the 1940 spring semester and enlisted in the Army. He began pilot training that September.

Ty was a second lieutenant, a 21-year-old with shiny new pilot wings, when he arrived in the Philippines in November 1941. War was raging in Europe, and Japan had been annexing territory in Asia for years. Like thousands of other soldiers in the Philippines, however, Ty saw those threats as distant from him as he was from home in the tiny town of Hyyannis, Nebraska, where the family had moved in 1939. Ty expected to spend two years serving his country in the tropical paradise.[1]

Ty and the twelve hundred other men of the 27th Bombardment Group, Light, were sent to Manila as part of a buildup of forces that had begun in 1935. President Franklin D. Roosevelt and his advisers had hoped a strong Pacific presence would deter Japan from attacking U.S. and European outposts and colonies. But Roosevelt's calculations were wrong. Despite the buildup, or because of it, Japan attacked Pearl Harbor on December 7, 1941, drawing the United States into World War II.

While the attack on Pearl Harbor is widely known, fewer recall the bombings hours later of other U.S. Pacific bases, including those in the Philippine Islands, an American commonwealth since the end of the Spanish-American War in 1902. Ty had been there for less than a month.[2]

Ty's parents, Hans Madsen Kokjer Jr. and Charlotte "Shy" Cobb Kokjer, were in near constant communication with their son from the time he left home until the war in the Pacific began. They kept 250 letters Ty wrote in 1940 and 1941 while he was in pilot training, along with some from college and earlier, as well as the handful that got out of Manila before war curtailed mail service. Ty wrote thousands of words after that, creating a diary of his war experiences in school essay notebooks. Each was in the form of a letter to his parents, signed, as were nearly all of his other letters, "Your loving son, Ty." His parents also saved three dozen letters they sent to Ty in the Philippines—and a few to Australia—that were returned once the war started. They never opened them. I did so seventy-five years later.

Ty wrote the diary while he was in hiding with a pro-American Filipino family after the United States surrendered the Philippines in spring 1942. At that time, an estimated seventy-eight thousand soldiers—twelve thousand of them Americans—were marched sixty miles up the Bataan Peninsula toward a prisoner-of-war camp in tropical heat in what would become known as the Bataan Death March. They were rarely fed, and Japanese guards were quick to shoot or bayonet soldiers who stepped out of line to fill their canteens at abundant water sources. Ty was one of the few soldiers to escape during the march. After eight months in hiding with the family, Ty realized he was likely to be recaptured. He buried all but one of the notebooks in a tin box, making sure one of the family members knew where it was if he couldn't come back.

In March 1945, at the end of fighting on Luzon Island, U.S. soldiers were stationed in Guagua, the market town near Ty's hiding spot. The Filipino family who protected Ty hadn't seen or heard from him for more than two years, and it's likely the family turned the notebooks over to the Americans there. Shortly after Japan surrendered in September 1945, a package containing the diary arrived at Ty's parents' new home in Lincoln, Nebraska. By then, Ty's family was fairly certain he had been sent to a prisoner-of-war camp in Japan. Now they hoped their only child had been rescued and was on his way back to them.[3]

That dream was short-lived. A few days after the diary arrived, they learned Ty had died months earlier. Hans and Shy solicited information from soldiers who survived the war, filling in the gaps from months of not knowing where Ty was or what he was doing. Histories and memoirs, published shortly after the war and later, added more detail. Those publications make it possible to connect the dots of Ty's life after his recapture because he was in some of the same places at the same time as the other soldiers and likely had similar experiences.

The letters and other documents spent forty years in a brown leather Samsonite suitcase lined with cotton in my aunt and uncle's cool, dry basement, which preserved them well. The original diary fared less well. Ty wrote in pencil in school examination notebooks, which had stiff paper covers and lined newsprint pages. The buried notebooks were subjected to heat and water damage during the rainy season. As a result, the writing at times is barely legible. Under different circumstances, that may have been a significant hindrance to understanding Ty's life in the Philippines. However, early in my work on this book, out of curiosity, I searched Ty's name online. To my surprise—and to my family's—one of the results that popped up was a link to a transcript of his diary at the National Archives and Records Administration in College Park, Maryland, not far from my home in Washington, D.C. Our family didn't know anyone else had a copy. We learned the Army had transcribed the diary, presumably searching for classified information or clues about missing soldiers, before returning it to the family. The National Archives said it likely received the transcript from the Army as part of a trove of war-related documents.[4]

The sixth notebook, the one he did not bury, is in much worse condition than the others. The police chief who turned Ty over to Japanese soldiers under duress was an American sympathizer. He hid the last notebook and sent it directly to Ty's parents, although it's not clear when it arrived. I have a hint to the chief's name, and Guill Ramos, a researcher in Manila, has found leads that might link us to his descendants. She found members of the Jose family that helped Ty and other soldiers. Their stories could fill in some of the gaps in Ty's diary, including who his other helpers and friends were— some of whom are identified only by their first names—and if Japanese forces punished them for their generosity.

Janet Kokjer Sothan, my mother's sister and the family genealogist, donated the notebooks, letters, photographs, videos, and his mother's wartime scrapbook—meant to be a gift to Ty after the war—to the Veterans History Project at the Library of Congress. The VHP collects and preserves stories of individual veterans from 1912 to the present. Ty's papers have been sorted and stabilized in archival storage. Researchers may see them at the Library of Congress, and someday they will be available on the project's website.[5]

Soon after Hans and Shy received Ty's diary, they also had it transcribed. The diary includes three gaps. Although the transcript at the National Archives contains a few pages not in the family transcript, its version includes the same gaps and ends November 18, 1942, at the end of the fifth notebook. The most likely explanation for the gaps is that either Ty ran out of paper or the books were destroyed by water. The typist of the family transcript wrote at one point the next few pages were consumed by mold or mildew, making them unreadable. At the beginning of one notebook after a gap, Ty wrote "(continued)," so it's clear he had written in a previous notebook. A third transcript, typed into a computer, is based on the second, with a few changes. Most are spelling corrections or spelled-out abbreviations: "Fort" rather than "Ft.," for example. The three gaps, all in 1942, are from August 14 to September 1, September 15 to September 25, and November 18 to December 9, although Ty backtracked to December 4 to start the next book. Ty had the last notebook with him when he was recaptured, and it concludes on that date, December 17, 1942.

The diaries and the many letters, written by Ty, his parents, and friends, are more like conversations we have today by email or text. Punctuation was casual or omitted. Some proper names are illegible, and some words are misspelled. My favorite mistake was made by a transcriber who didn't know the difference between a gorilla and a guerilla. Small primates exist in the Philippines, mainly macaques, but gorillas do not. There are other discrepancies, mainly event dates. Because the Philippines are across the International Dateline, dates reported from there versus dates reported in the States are off by a day. Memoirs by Death March survivors written shortly after, or even years after, the war pop up now and then online and in bookstores that sell used books. Many include dates that do not jibe with Ty's reporting. It's no surprise veterans could not always remember the exact dates some things happened, especially if they did not keep diaries.

Ty's friend Robert G. Bjoring wrote a memoir of his life— from his birth in Oslo, Minnesota, to his time in the Philippines and his postwar Air Force career, in which he rose to the rank of lieutenant colonel. The unpublished manuscript is in the files of the historian's office at Joint Base San Antonio–Randolph. Bjoring and Ty wrote about each other, but Bjoring's memoir contains few dates related to his war experiences. Some books have been published more recently as veterans die and their children find documents in attics and trunks. Ty's diary is a rare example of a contemporaneous account of his actions from the time his ship left San Francisco, the beginning of the war, through the Death March, and while he was on a boat with the Filipino police chief on his way to again becoming a prisoner of war. His is the only account I know of that tells the story of someone who spent so much time living with Filipinos.

Those who knew Ty well had died or were elderly, some with failing memories, when I began my research, so some of my questions about him and his family have gone unanswered. Hans died in 1969, and Shy in 1972, both when I was in college. I never got to talk to them about Ty.

We have only black-and-white pictures of Ty, meaning I don't know what color his hair was, though probably brown. Ty wrote at one point in his diary that he would have a better chance of blending into the local population if he was shorter, had a better tan, and had brown eyes. He apparently inherited the Kokjer family's blue eyes. A fire at the Military Personnel Records Center in St. Louis in 1973 destroyed millions of pages of documents dated from 1912 to 1964. Copies of records held by other agencies can sometimes reconstruct the official life of a soldier. For Ty, all the military researchers could find were his enlistment and discharge papers. Neither contains a description. I don't have a copy of his driver's license. His birth certificate doesn't even list his weight.

My mother, Phyllis Kokjer Beck, who died in 2019, pulled pictures she had of Ty and his parents from her many scrapbooks and enjoyed hearing about my research into his life.

Even with her failing memory, she recalled that Ty treated her as a pal, despite their six-year age difference. My aunt Janet, more than eleven years younger than Ty, has more memories of Hans and Shy. Janet has traced the Kokjer family to its roots in Denmark and elsewhere and kept records. As eleven of Ty's cousins gathered for Shy's funeral and a dispersal of her property, Janet said the only things she wanted were the letters, diaries, and scrapbooks. She has been an invaluable source of information and support for this book. She and her late husband, Norman L. Sothan, a U.S. Navy commander and aviator, guarded the diary and the letters after Hans's and Shy's deaths.

Racism was a presence on both sides before and during the war, in culture generally and in propaganda. Americans regarded Japanese people as short, buck-toothed, coke-bottle-glasses-wearing people who were backward and incompetent. "Our own propaganda had us pretty well convinced we could whip the little myopic bastards in a couple of weeks when, not if, they attacked," one enlisted man said after the war. Some propaganda posters depicted Japanese soldiers as rats about to be caught in the mousetrap of American military might. In films, these characters were played by American actors in yellowface. On the other side of the ocean, Japanese citizens regarded American soldiers—and their country—as poorly educated, poorly trained, soft and degenerate, without honor, and filled with gangsters like those in movies from the 1930s exported to Japan. Japanese soldiers thought even less of Filipino soldiers, most of whom lacked weapons and some of whom wore canvas shoes and helmets made of coconut shells. Americans were astounded that Japanese soldiers would kill themselves in the emperor's name rather than surrender; Japanese soldiers were equally astounded that Americans would dishonor themselves and their country by surrendering rather than die. Many American soldiers and veterans, Ty included, referred to Japanese soldiers or the country's residents as "Japs," a term that is and was racist. I have used the word only in direct quotes when it is important in context. Ty occasionally used other phrases that were likely commonly accepted by some in the 1940s while considered racist by others, as they are today. They remain in his diary and letters, but I have not included them in the text. I have, however, included some of his thoughts about race and racism.[6]

Like some others who wrote about their war experiences, Ty wished to spare his parents from the worst things that happened to him. He said he would tell them after the war. He also masked the identities of Filipinos who helped him, although there are clues, which Guill Ramos has used to find people who remember families who hid Americans during the war. My plans to join her in the Philippines in 2020, 2021, and 2022 fell victim to the Covid-19 pandemic. I was finally able to make the trip in April 2023. But Covid caught me toward the end of my two-week tour, and I isolated in my hotel room instead of visiting the families. Guill did so did so after I returned home. She interviewed two elderly sisters who were young children during the war. They remember their father's generosity toward American soldiers and their fear when Japanese soldiers knocked on the door looking for Americans. But they have no specific memories of Ty.

Through my research I have learned the names and have information about the fates of some of Ty's friends. When his parents visited the Philippines in 1958, they met with Dr. Jose Jose, who had befriended Ty, according to a family history. He is one of the Filipinos in a picture taken at Ty's grave. He was the father of the two women Guill interviewed. We believe his family was Ty's primary host when he was in hiding.

Based on Ty's diary, it seems unlikely he ever fired a shot in combat. Whether he was a hero or the family built him up as one is a question my uncle, a Navy combat veteran, raised.

Ty performed tasks behind the lines, and sometimes at the front line, that enabled soldiers trained in ground combat to perform their duties. To me, all the soldiers and nurses who volunteered for duty are heroes. Without them, it would have taken longer to win the war, and more people would have died.

I hope Ty's story will add to the knowledge about soldiers like him who sacrificed to keep the United States free. The large family that hid Ty over eight months, and others in their small community near Guagua, in the Pampanga Province, are almost all identified only by their first names or Americanized nicknames. They were not the only Filipinos who endangered themselves to help a U.S. soldier, and in that way, their anonymity is fitting. The family that hid Ty and two other Americans represent the many Filipinos who helped soldiers return to their families and allowed our family to learn more about a lost cousin.

Ty and his classmates at Randolph Air Field were photographed in their flying gear for their graduation photos, which appeared in a yearbook called The Gig Sheet. A white silk scarf is tucked under his jacket. Photograph courtesy of Joint Bast San Antonio-Randolph

Chapter One

A "Swell" Small-Town Life

"Pompous and grandiloquent / The leader of our band so eloquent."

—Caption under Ty Kokjer's photo in the Kearney High School yearbook, 1937

Like many of the young men who would fight in World War II, Madsen "Ty" Cobb Kokjer grew up in a small town, almost oblivious to world events that would shape his life. He was born December 23, 1919, in Kearney, Nebraska, the only child of Hans Madsen Kokjer Jr. and Charlotte "Shy" Cobb Kokjer.

Ty and his family were as square as they came, when square meant honest, respectable, and perhaps, naïve. His Danish grandfather, who immigrated to the United States at age 17 in 1874, passed to his children and grandchildren beliefs in education, temperance, hard work, duty, and patriotism.

Kearney, the seat of Buffalo County, is on the north bank of the Platte River nearly two hundred miles west of Omaha. By the time Fort Kearny was founded in 1848, the American bison was nearly extinct, but thirty thousand pioneers passed through on their way west the next year. The Pony Express, the Mormon and Oregon Trails, the railroad—and much later, Interstate 80—all followed the river. From the first homestead in 1871, the town's population grew to 9,600 in 1940. By 2020, Kearney (pronounced KAR-nee) was a college town of 34,000 residents. Thousands of people flock there in March to see the migrating Sandhill cranes.[7]

Ty graduated from Kearney High School in 1937. The family moved to Hyannis, more than three hours northwest of Kearney, in 1939, while Ty was in college. He wrote in his wartime diary that July 17, 1942, marked three years since they had taken over the Hotel Hyannis. The next year, he said, they paid off "Bry" and the hotel was theirs. The change in management rated a two-sentence, front-page article in the *Grant County Tribune*, July 12, 1939: "NEW PROPRIETORS FOR HOTEL HYANNIS / Mr. and Mrs. Leonard Breuhaus have disposed of their interests in the Hotel Hyannis to H.M. Kokjer of Kearney. The new proprietor is expected to take possession on July 17." The move took them from farming country to a rail stop in cattle country. Hyannis, the seat of Grant County, is deep in the state's Sandhills, where State Routes 2 and 61 cross. The town's population peaked at 449 in 1940; by 2020 it was 165. Of Nebraska's ninety-three counties, Grant tied for eighty-eighth in population in 2020, with 722 residents, down from 1,327 in 1940.[8]

Considering that the family lived in Hyannis for a far shorter time than they lived in Kearney, it's surprising Ty thought of it as home and as part of his future. But Ty and his parents had friends at two ranches near Hyannis before moving there, including members of the Abbott family. That ranch, 40 miles north of Hyannis in Cherry County, is big enough to show up in an online map search. In June 1935, when Ty was 15, he worked on a ranch near Hyannis. He didn't name it, but the Abbott ranch seems likely. He earned $45 a month, plus an extra $5 to get up early and wrangle the horses. Room and board were part of the deal. He lived with two other ranch hands, all there for the haying season. "We really have it swell in our bunk car," he wrote to his parents. He wondered about their plans for the July 4 holiday and said he could spend the day with his friend Marvin Metzger's family on their nearby ranch, where he worked the summers of 1937 and 1938. He also praised the cook: "I just had to tell you about the dinner we had this noon. Baked chicken that was juicy inside with the skin well browned on the outside, mashed potatoes, gravy, dressing that was also moist but had a certain crispness, ice tea, and cream pie with a real flaky crust. She can really cook."[9]

Ty's cousins, Janet Kokjer Sothan, more than eleven years younger than Ty, and her sister, Phyllis, six years younger than Ty, remembered that he was in the high school marching band and was its drum major. Janet recalled that he once threw his baton into the air, only to see it land in a tuba. In his high school senior yearbook, the slogan under his picture reads: "Pompous and grandiloquent / The leader of our band so eloquent." He was active in athletics and other clubs, and by all accounts was a friendly, outgoing boy. After the family moved to Hyannis, Ty visited Kearney and his friends there whenever he could, but he wrote that, by the time he finished Army training, many of them had left or were at different stages in their lives—working, married, and with young children.

During high school, Ty got into some sort of trouble. His parents sent him to the Pillsbury Military Academy in Owatonna, Minnesota, for a year. Neither Janet nor Phyllis could remember what prompted their decision or how they paid for the private school during the Great Depression. The school, now called Pillsbury College Prep, is no longer a military academy and has no records from the 1930s. Punishment or not, Ty remembered his time there fondly and visited the campus when he was in Minneapolis for a college fraternity convention.

No letters survive from Ty's time at Pillsbury. The earliest letter his parents saved is four pages he wrote to his grandmother, Augusta "Nanny" Cobb, in April 1932. When long-distance phone calls were expensive, families stayed in touch through letters, at three cents for a stamp, and penny postcards. Ty's penmanship was good, but even though his team won the fourth-grade spelling contest at school, that skill wasn't always reflected in his letters. Spring had arrived. He told his grandmother he planned to take his bike out of storage, and robins in the yard were making "a tarrable chatter." He had been playing with 129 of his countless toy soldiers on the living room floor, something he recalled and wrote about during the war. "I am getting along quiet well in scouting," he wrote to his grandmother. He was the bugler for his Boy Scout troop, and the scouts had recently cooked dinner over a campfire on a hike led by his mother, which qualified him for a merit badge.[10]

Ty's fondest memories were of the few years his family ran the Hotel Hyannis. They lived in two rooms on the third floor connected by a bathroom and spent time in the hotel's first-floor lounge, where they could greet guests and keep an eye on the help. Once Ty was overseas, Hans and Shy did what many parents do when their children leave home—they converted his bedroom into a sitting room. The hotel, which opened as the DeFair in 1898, is a national historic landmark. When Hans and Shy ran the hotel, it was owned by a committee of ranchers who wanted to ensure there was a place for cattle buyers to stay. Hans and Shy occasionally wrote about selling the hotel, so they had a contract or a stake in the business, but county land records show they didn't own the property.[11]

Phyllis remembered how nice Ty was to include her in activities with his friends when they visited Hyannis instead of leaving her behind with the kids, including Janet, who had a fine time with cousins closer to her own age. Their family—Emerson, Hans's youngest brother by a decade, and his wife, Winifred, and the two girls—lived in Wahoo, Nebraska, 150 miles from Kearney and 300 miles from Hyannis, so their visits were infrequent. Hans and Shy saw more of his brother Tom's family in Sidney, Nebraska, about 140 miles west of Hyannis. Shy was close to Tom's wife, Isabel, or "Izzy." They often saw relatives on Shy's side of the family. Her father, U.S. Cobb, lived in Funk, Nebraska, a few miles south of Kearney, in the house where she grew up. Shy's brother, Robert, his wife, Grace, and their family—Chandler "Chan," Robert "Bobby," and Margaret—lived in Alliance, sixty miles west of Hyannis. The Cobbs moved to Hastings, an hour east of Kearney, during the war, where Robert found work at an ordnance plant.

Ty's father was born in 1888 and attended the University of Nebraska. During World War I, he served in the Army at Fort Riley, Kansas, in the commissary department. After the war, he had a number of jobs. He was a steward at the Nebraska Industrial School, a reform school, and later worked at the Nebraska State Hospital for Tuberculosis, probably in an administrative role. He was a traveling salesman for a wholesale grocery business and for a janitorial supplies company. Shy, born in 1897, was a volunteer aide at the hospital, visiting with patients and delivering flowers or gifts to their rooms.[12]

After Ty joined the military, Hans and Shy often shared news with their son about their days. Their missives paint a clear picture of what it was like to run the business that continued to include Ty's name on the hotel stationery as an owner. The letters they sent while Ty was in the Philippines were returned when air mail was suspended shortly before the war started. About three dozen survived. The first, dated December 3, 1941, acknowledged the first letter Ty sent from Manila and called it "a precious jewel." Their letters are full of news about ranchers and townspeople and the problems of keeping good help when higher-paying jobs became available as many young men left for the war and women had more opportunities, including at war production plants.

The first half of February 1942, they wrote, was worse than the same period in 1941. During "the regular winter lull," Hans reported eleven rooms out, four of them to insurance men who would stay the week. He said the hotel averaged $12 a day in rooms and $30 per day on meals in the first part of 1942. With prices unchanged from 1940, total sales volume was $21,000, with profits on meals 8 percent to 10 percent lower than the previous year. Hans was happy when the snow melted and ranchers came in more often to eat. After a dance in town, the restaurant did more business than any day since the beginning of the year. In her letter the next day, Shy wrote that they were tired: "We were all up late last night selling liquor."

Over the winter, they debated whether to let the waitress go, but feared another employee might quit soon. One waitress, "our ... redhead," wasn't fast enough for the busy summer season. (There are never any full names in these accounts.) When help was short, Hans and Shy took on extra chores themselves, cleaning rooms, cooking, and waiting tables. Shy wrote in mid-February 1942, "Today, I am the cook. Oscar is 'gone with the wind.'" He apparently drank too much, and he and Bessie went to work on a ranch where they could save money. After Oscar left, Hans cut the steaks. He reported that a customer said their T-bone was better, and less expensive, than one he had eaten recently at Omaha's finest hotel. A few weeks later, Hans was still cutting meat after a new man in the kitchen made a steak mistake: "The other day, he stripped all the fat off the T-bones." Another letter from Shy indicates her deep involvement in running the hotel. Snowbound in April 1942, with Hans out of town and paying guests who couldn't leave, she took care of things on her own.[13]

Both Shy and Hans made frequent trips to nearby towns to buy chickens and eggs and to recruit help. In March, Shy reported "three new girls" were on board. On a trip lasting several days, she hired two young women from Merna and one from Mason City, both more than one hundred miles east of Hyannis with about the same population. "I just worked the little towns like a traveling man," she wrote. Shy joined several clubs, which often met at the hotel. When the Eastern Star, a Masonic organization, and her book club met at the hotel in the same week, she spent more time waiting tables than joining in.

As Ty started classes at the University of Nebraska in Lincoln in the fall of 1937, he wrote several letters to Hans, and later to Shy, about the fraternity he wanted to join. Shy said the Sigma Chi house had a bad reputation because members drank too much. "I don't know why but I admit I feel mother won't want to see me or have anything to do with me from the way she writes, yet I know it is silly to feel that way because she is only saying what she thinks is best," Ty wrote to his father, assuring him that only a few members drank and they weren't often at the house. "But one thing is for certain, if mother does not change her mind about this fraternity and will not be happy with me here, I will break my pledge because I want her to be happy." He said he assumed she would eventually concur in his judgment that the Sigma Chis were "a swell bunch of boys." Shy did come around, perhaps on the promise the fraternity would serenade her as the "Sweetheart of Sigma Chi." The family still has his fraternity pin.[14]

Ty planned to study medicine and belonged to NU-Meds, a premedical social group, according to the 1938 Cornhusker yearbook. By his sophomore year, according to the 1939 Cornhusker, he was no longer involved in that group, but he was a twirler in the varsity marching band; a newsletter published while he was in flight training said Ty could twirl three batons at once. He also participated in ROTC (Reserve Officers' Training Corps), something male students were encouraged to do and sometimes required to take part in at land-grant universities such as Nebraska. In the fall of his junior year, Ty withdrew from school for surgery to have his appendix removed. He returned in January 1940 for the spring semester. His letters home tell of his grades, dates, movies, and his constant struggle to stay on budget, including what he confessed was too much money spent for cigarettes. But he was distracted.[15]

Ty had never been to the East Coast or overseas. He wanted to have a little adventure before settling down to marry, raise a family, and make his way in the world. "Well something happened this afternoon that I have thought of and have been afraid to hope for. I took and passed the examination for Pensacola the Naval Academy of the Air," Ty wrote his parents May 7. He was one of five to qualify of fifty who took the test. Ty went to Kansas City for more tests and interviews, then wrote, "It looks as though I am going to make it if you have no objections." When he described the chance to become a pilot in a letter to his parents, he signed off "Your loving son and future aviator (I hope)." Ty visited Naval Air Station Pensacola in Florida and brought back a shell as a souvenir.

Soon, Ty also qualified to train as an Army pilot. "And I sure can't decide which one [Army or Navy] would be the best as you hear one thing from one person and then another thing from the next one." He asked his Uncle Tom Kokjer in a May 10, 1940, letter about the choice. Tom was a World War I pilot and flight instructor who trained at the Army's Kelly Field near San Antonio, Texas. He was so seriously injured in a plane crash that he was never sent overseas and missed his mother's funeral in 1920. If Tom replied to Ty, the letter no longer exists. But Ty's respect for his uncle likely played a role in his choice of the Army instead of the Navy.[16]

In retrospect, events from the 1930s to early 1941 should have alarmed most Americans. In 1931, Japan attacked Manchuria, in the northeast area of China, a strategic asset and a source for scarce raw materials. The country's main goal, announced in late 1938, was to create and lead a "new world order in East Asia." It assumed other Asian countries would reject their status as Western colonies and join the Greater East Asia Co-Prosperity Sphere. Meanwhile, Germany marched into Austria in March 1938 and invaded Poland in September 1939, which led Great Britain and France to declare war on Germany two days later.[17]

In his 1940 State of the Union speech, President Franklin D. Roosevelt reiterated his promises to keep the United States out of war, while urging the country to avoid becoming "American ostriches": "There is a vast difference between keeping out of war and pretending that war is none of our business."[18]

Despite these events, neither Ty nor his parents wrote that his decision to enlist was influenced by a possible war or that he would fight in one. In the summer of 1940, before he entered flight training, Ty was at ROTC training camp, and he did bring it up: "Say this war business is really popping up all over. I sure didn't expect the Danes and the Norwegians to get into it. I'd sure hate for us to get into it now, but if we do I'm glad I am in the position I'm in. Everyone here is pretty pessimistic now. They seem to think we will be in it before the summer is over. However, I doubt it."

Denmark, where the family still had cousins, and Norway fell to Germany in April 1940. Perhaps Ty and his family discussed the war in their Sunday afternoon telephone calls or when they were together. His father and his Kokjer uncles, Emerson and Thomas, were World War I veterans. Hans served in the United States; Emerson deployed to Hawaii and France with the Signal Corps; Tom, as noted, was a pilot who trained at Kelly Field and became a flight instructor there. Roy, who worked for the Union Pacific Railroad, was likely considered more valuable at home than on the front lines.[19]

The next step in Ty's military career was a trip in August with Hans to the recruiting office in Scottsbluff, in far western Nebraska, where Ty enlisted in the Army. Afterward, they drove seventy-five miles to Tom's house in Sidney, where cousin Bobby Cobb and Shy were waiting for them. "Then we took in the [county] fair that evening and a good time was had by all," Ty wrote about the momentous day in his war diary. "Well, two years ago today, Dad ... you were present when I enlisted in the Army." Shy wrote later in her scrapbook, "Daddy saw you take your oath to your country. He said he realized keenly that he had given you to our country." The scrapbook was to be a gift to Ty after the war and is filled with his mother's writings, letters, telegrams, newspaper clippings, family pictures, and notices of friends' and relatives' weddings and babies.

The railroad was a constant in American life in the 1940s. It was punctual enough that Hans and Shy often closed their letters by saying the train would be in soon to take the mail. Hans, who had a reputation as a wit, wrote to Ty in June 1942 that they had paid his $3 fraternity alumni dues. "I thought you would want to keep that up. The train just came in with the morning's papers, so I will get ours and see if I can find you in the funnies." Trains haven't stopped in Hyannis for years. Now they speed through, most carrying coal east from Wyoming, with a piercing whistle at the Route 61 grade-level crossing four blocks west of the hotel.

To feed, clothe, and arm soldiers after the war started, the government rationed sugar, gasoline, shoes, metal, fuel oil, coal, firewood, nylon, and silk. Large portions of other materials, such as rubber, were under Japan's control, leading to rationing of tires for cars and bicycles. Factories that made cars switched to making jeeps, ambulances, and tanks. Rationing affected the hotel. It cut into the meal business from ranchers, working without enough help, and there were fewer travelers on State Route 2, a main east-west highway, to book rooms.[20]

Across the country, there were drives to collect rubber, tin, and other metals for the war effort. Farmers donated old plows, women donated cooking pots, children gave metal toys, and some donated the bumpers from their cars. The drives, promoted by movie stars and local publishers, created a patriotic camaraderie and made people at home feel as though they were contributing to the war effort.

Shy pasted a newspaper story about the drive in her scrapbook, noting that Grant County held the "undisputed lead in the contest," collecting 111.53 pounds of scrap metal per person in the 1942 contest sponsored by the owner of the *Omaha World Herald.* The county won a $1,000 war bond. Nationally, Nebraska came in sixth. A paper written in 2007 for the National Bureau of Economic Research said Nebraska had "the most successful state drive yet, [and] the Nebraska model was widely copied." But the author, Hugh Rockoff, an economics professor at Rutgers University, also wrote that the drives were more important as morale boosters than for the amount of materials they collected.[21]

Business at the hotel, which rarely boomed, slowed during the war, and keeping employees became even more difficult. In the fall of 1944, the Kokjers left the hotel business. The following spring, they moved to an apartment half a block from the capitol in Lincoln, where Hans was an inspector for the state Department of Labor. There, as they had in Hyannis, they waited each day for news from their son, Lieutenant Madsen C. Kokjer, of the U.S. Army Air Forces, stationed in the Philippines.

The Kokjer family lived in Clarks, Nebraska, where Hans settled after spending a few years in Iowa. Standing, from left: Thomas E., Ralph Leroy, Meta Elizabeth, and Hans Madsen Jr. Seated from left: Malina Alice, H. Emerson, and Hans Madsen Sr. The photo was probably taken in about 1901. Family photo

The Cobb family, circa 1905-1907. Charlotte "Shy" is in front. From left: her brother, Robert, her father Ulysses S., and mother Augusta. Family photo

Ty was about a year old in this photograph. Born December 23, 1919, in Kearney, Nebraska, he was an only child and wrote in his war diary that he knew his parents had sacrificed for him. Family photo

Chapter Two

A Flight Cadet's Adventures

"It really feels swell to be up there by myself."

—Ty Kokjer, writing about his first solo flight, October 1, 1940

On the morning of September 6, 1940, Hans, Shy, Uncle Tom, and Aunt Izzy saw Ty off on the train to San Francisco. After visiting his Uncle Roy Kokjer's family in Larkspur for a few days, Ty flew to Los Angeles. He sent his parents a United Air Lines postcard with an image of high-flying fine dining and praised the short flight. He reported for duty at Oxnard Army Air Field, a dusty spot near the coast about halfway between Malibu and Santa Barbara, where he was one of forty-six cadets in class 41-C. The Lincoln, Nebraska, airport, where Ty had expected to begin his training, and Cal-Aero Training in Oxnard were two of the fifty-six civilian pilot training schools around the country under contract with the government to train the large numbers of pilots, navigators, and meteorologists who would be needed if the country went to war. Congress authorized a plan to take the Army Air Corps from the twenty thousand personnel it had in 1938 to one hundred fifty thousand by June 1941.[22]

From the time he left home, Ty wrote to his parents nearly every day, skipping a day if he was on leave in Los Angeles, when he was hospitalized with a broken leg, or when he ran out of paper or stamps. Hans and Shy had saved a few letters Ty wrote when he was working on ranches during the summer and from his time in college. But from September 1940 until the war interrupted mail service overseas, they saved every word he wrote: 250 letters. Ty saved none of his parents' letters until he was deployed, when he kept their last two letters in his shirt pocket. He had good grades in penmanship in school, and even the letters he wrote in pencil on cheap paper are legible. Spelling and punctuation were somewhat random. But these daily jottings weren't meant to be formal essays; they were his end of a constant conversation with his parents about who he met, what he was learning, how he was often short of money, memories of family, events, and when he might get home on leave.

At Oxnard, Ty was issued his "flying clothes": coveralls, a wool sweater, a fur-lined helmet, goggles, and a silk scarf. "The food is swell," he wrote, using his favorite adjective: the "awesome" of the era. On the other hand, "This place is stricter than the dickens," with demerits for those who violated the rules. Training began early September 14 with two hours of drills in the morning and two more in the afternoon. Three days later, he flew with his instructor. They went out over the Pacific Ocean, and the instructor pointed out landmarks so Ty wouldn't get lost when he was on his own. The instructor, a local police officer, demonstrated maneuvers, including a spin, so Ty would know what they felt like. He praised the training plane, an open-cockpit, Stearman PT-13 Kaydet biplane with a 225-horsepower engine. Stearman, later acquired by Boeing, manufactured thousands of training planes in the lead-up to the war. "It really is a thrill to fly planes, and I can see that it is going to take a lot of concentration." On this first day, he was allowed to taxi back to the line, set the brakes, and shut off the engine. He described the day: "It is really swell."[23]

By the next day, he was sharing controls with his instructor as they landed the plane. "That throws one of my biggest obstacles out of the way." Over the following days, he was banking the plane at a forty-five-degree angle, gliding, following a straight course, and ascending into clouds, "big billowy things," at fifteen hundred feet. "My instructor is the cool type of fellow. He doesn't get irritated when you do something wrong. He corrects you and shows or tells you how to avoid it happening again." That was lucky. A few days later, "I was recovering from a stall and just about did an outside loop and almost lost my instructor out of the plane. He didn't have his safety belt very tight and he left his seat about four inches. However, there was no real danger of losing him. I really would have been on my own if it had happened." In a few more days, he was able to take off, ascend to four thousand feet, above the clouds, and land in a crosswind. "It really is fun." Soon after that, Ty flew with a check instructor. "It made me nervous and as a result, I did everything wrong."

Cadets at Oxnard learned basic flight skills and took classes in math, airplane mechanics, navigation, and meteorology. Before moving to a military base for advanced flight and officer training after Thanksgiving, they needed to pass their classes and a check flight. Of those who had never flown before, Ty was the second in his class to solo, on October 1. "It felt kind of funny at first," he wrote, adding, "it really feels swell to be up there by myself." Three days later, the first of the forty-six cadets washed out. In essence, he flunked flying school. Another was to take a "wash ride" the next morning. "He just couldn't seem to get the hang of landing a plane, and they were afraid to let him go up alone," Ty wrote about the second pilot. Another cadet damaged a plane during a landing. "As the instructor said today, you have to fly a plane 'til it is stopped." The Army expected an overall washout rate of 50 percent. More selective recruiting and demanding instruction kept washout rates high, but it raised flying competency and lowered the number of plane crashes later.[24]

Ty wrote to his parents how much time he had flown solo to the minute and what new maneuvers he was learning. Over the next weeks, Ty began to practice what he called acrobatics (more often called aerobatics), learned to fly in hot weather and updrafts, and practiced forced landings. Then came more complicated maneuvers: chandelles (a series of 180-degree turns), pylon eights (figure eights based on marks on the ground), stalls, spins, vertical reverses (ascending until the plane loses speed, then reversing into a steep dive), and half rolls. "I about tore myself in half trying to do them solo today." And there were snap rolls and the Immelmann, a combined quick turn and ascent. "Gosh, I don't see how I am ever going to do these snap rolls as I couldn't tell where I was going today," he wrote, adding that he expected to catch on eventually. He flew to seven thousand five hundred feet, where it was frigid in the open cockpit. At that height, above the clouds, he could see Los Angeles and Catalina Island, seventy miles away. Later, he flew to ten thousand feet, where he and his instructor took turns doing aerobatics.

Los Angeles was more than a vision from aloft. Marvin Metzger, a friend from Gordon, Nebraska, was a graduate student in art there. He enlisted in the Navy in 1942. After the war, he returned to run the family ranch, where Ty had worked two summers. When Ty wasn't the duty officer or confined to the base because of demerits, he went downtown for frenetic weekends. One October weekend, he and Metzger ordered the special sixty-cent menu at a Chinese restaurant, which Ty called "a very fine place to eat," noting it had a pond "where you can fish out your own dinner." After a visit to a night bazaar, they went bowling and saw the movie *Boom*

Town starring Spencer Tracy, Clark Gable, Claudette Colbert, Hedy Lamar, and Chill Wills in a story about oil tycoons and a love triangle. The next day he ate a picnic lunch with relatives. Another weekend, he and Metzger split dinners of fried shrimp and chow mien, followed by a floor show and dancing until 9:30 p.m., miniature golf until midnight, and fried oyster sandwiches at a drive-in restaurant until 1:30 a.m. Ty went to Hollywood to see Radio City NBC, the network's West Coast broadcast hub. At the gaudy Egyptian Theatre (built by Sid Grauman)—"they have live monkeys in front"—he took in a movie, *Pride and Prejudice*, starring Greer Garson and Laurence Olivier. He walked by Grauman's Chinese Theater.[25]

"Well, I am just back from a really swell weekend," he wrote to his parents October 27. It included a 10:30 p.m. dinner at a steakhouse, with perhaps the only cut of beef he ever praised that didn't come from Nebraska. "It was as tender as ours, but not quite the flavor." The next night, he, Metzger, their dates, and another couple sat at a restaurant table next to the dance floor. It was the quintessential midcentury, fancy dinner: shrimp cocktail, chicken broth, celery stuffed with cheese, hot rolls, green peas, and string beans. Ty had half a lobster in the shell as the others ate steaks, and everyone had salads and peach parfait. They danced between courses until midnight. Afterward, they went to Hollywood and peeked into some famous clubs, including the Brown Derby. "This weekend really cost me, but it was sure worth it," he wrote. Another time, he and Metzger went to the port of San Pedro, where they saw part of the U.S. Navy fleet. He saw his first ocean sunset. Another time, he went to the beach because he heard the waves were high. While he was in a new place, he wanted to do things he couldn't do elsewhere.

In 1940, as Ty learned to fly, some Americans began to worry about war with Japan and Germany, despite one poll that found 70 percent of Americans were opposed to war. Japan's population of 80 million was growing by a million per year, too many to feed without goods from other countries. Manchuria, which today is split between Russia and China, just across the Sea of Japan, could supply Japan's needs. Fighting began there in September 1931, and Japan annexed the territory in 1934. In Europe, the situation was dire. On June 4, 1940, Great Britain completed the evacuation from Dunkirk in France of more than three hundred thousand Allied soldiers backed up to the English Channel. Italy joined forces with Germany on June 10, and France surrendered to Germany on June 21. In September, Japan, Germany, and Italy signed an alliance known as the Tripartite Pact. Congress passed the Selective Service Act the same month, allowing men to be drafted into the Army during peacetime, but continued to forbid trade with or aid to warring countries under the 1936 Neutrality Act.[26]

At the halfway point in his training, in mid-October, with twenty-six hours of solo flying, Ty wrote his parents, "Another week is gone and I'm still in this man's army and sure hope I can stick it out." He felt he didn't yet know much about flying. Two weeks later, he and his instructor had a hard time landing in a thirty-mile-per-hour wind. When they got on the ground, "there were about 10 fellows to grab the plane so that it wouldn't upset. It was fun though." Once he accumulated thirty hours of flying, on October 30, Ty went up with a check pilot, who had few corrections. "I really did swell. ... He said I gave him the best ride any student he has ever had had given him. He couldn't get over how nicely I had done." Ty passed another check flight in November.

Like most of his Republican relatives, he approved of the election results that November, at least at the state level. Dwight Griswold, a Republican newspaper owner from Gordon, was elected governor, and Wendell Willkie beat President Roosevelt 57 percent to 43 percent. Nationally, however, Roosevelt won an unprecedented third term with 55 percent of the popular vote and 85 percent of the Electoral College votes. Ty was headed to Los Angeles for a final weekend with his California buddies.[27]

By mid-November, as Ty prepared to leave Oxnard, the instructor told Ty he should have no trouble getting through advanced training and earning his commission. After that, Ty relaxed a little. "I lay out in the sun and got a nice sun bath this morning and thought how you must be enjoying the snow and cold there in Nebraska," he wrote to his parents. Ty turned in his flying equipment and worried that his weight, 187 pounds, a result of all the good food, would make him too heavy to fly. He was right. The weight limit for fighter pilots was 160, although those flying bigger planes could weigh a few pounds more. He was warned again about his weight at his next post, and he changed his eating habits to lose weight.[28]

To celebrate the end of training, there was a dance, his first Army social function, with young women from Ventura College selected by the dean.

His last letter from Oxnard, dated November 19, 1940, was a quick note to his parents about the farewell dinner with his instructor and a few other cadets. He was on his way to Texas and the second phase of flight training. He had told his grandfather to expect him home for Thanksgiving as a surprise, but now, he told his parents, with only five days before he was to report to his next assignment, he wouldn't have time.

Ty, flanked by his parents, Hans and Shy, appears to be
about 18 months old. His father worked a number of jobs,
and Ty missed him when he was a traveling salesman. His
mother led his Boy Scout troop, belonged to several clubs,
and did volunteer work. Family photo

Ty did well in grade school in Kearney, Nebraska, including
being on a team that won a spelling contest, as a clown in a
class play, and as a Boy Scout. Family photo

Ty got into trouble as a teenager. His parents sent him to
Pillsbury Military Academy for a year. Ty wrote about his
time there fondly and visited the Minnesota campus when
he was in college. Family photo

Chapter Three

Really in the Army Now

"We are going to live a dog's life around here for the next five weeks, but I sure like it."

—Ty Kokjer, Randolph Field, November 25, 1940

If Ty thought things were strict at Oxnard, discipline at Randolph Field, northeast of San Antonio, Texas, caught him off guard. Now, he was really in the Army. As World War II loomed, Randolph, with its art deco tower topped by shiny blue-and-gold tiles, became the largest basic flying school in the United States.[29]

Ty made a quick, surprise visit home for Thanksgiving after all, then traveled to Randolph, arriving at 2 a.m. on November 25, 1940. After taking a nap, he was on the go, learning how to stand at attention and march under orders from upperclassmen, "just like at Pillsbury, only this is much more impressive. ... We are going to live a dog's life around here for the next five weeks, but I sure like it." He submitted his orders and was assigned to a company, squad, and room. Supplies were issued. "There is a certain place for everything and a certain way for it to be folded. We had last evening to fix things up," he wrote on November 28. He tried out for the drum and bugle corps. Members didn't have to drill or take care of a rifle, a decision he may have regretted later. When he took gunnery training in March, he hit the target only once.

"I'm telling you, this is really an Army post and is as much like 'West Point' as any pictures you have seen. We don't walk, sit or stand any way but at attention. And we can't talk except when spoken to or else when we are in our rooms." The cadets had to look straight ahead at all times. "As yet, I haven't seen what the ceiling of the mess hall looks like. ... There is a lot of tradition about this place."

Planes flew from 7 a.m. to 11 p.m. The cadets got up at 5:45 a.m. Lights out was at 9:30 p.m. unless they were flying. Despite the strict rules, Ty wrote a few days later, "The more I am here, the more I like it." He liked the food, but he needed to lose a few pounds. He was adjusting to the climate. With Christmas a few weeks off, he wrote, "It doesn't seem like that time of year here with no coat or heavy clothing."

Ground school started first for the flying cadets, with classes in military customs, table manners, how to call on officers, and what to wear. "They are really preparing us to be officers and gentlemen." They would earn commissions as second lieutenants if they completed their training. At Randolph, they earned $75 a month, plus room, board, and uniforms. Once they were commissioned, their pay rose to $205.[30]

Now that the cadets knew how to fly, they were ready to move to a more complex plane, the new single-wing BT-14 with a closed cockpit (BT for basic trainer). "They are really swell ships. 450 horsepower. The cruising speed is between 125 and 140 miles an hour and they glide and climb at 90 mph, and land between 75 and 85 mph." The plane, he added, had "a whole slew of instruments to mess you up." It took Ty a while to figure it out, and he complained that he was turned around—north was west to him. "It really has me worried. However all I can do is try, and if I am not able to make the grade it is because I haven't the qualities to make the grade as a military pilot, but I sure am going to work. I bet I lose a pound every hour I am up in the air plane the way I am all wet with perspiration when I get down."[31]

By December 10, he was flying solo in the new plane, and the next day, cruising above the clouds, he calculated that if he could fly to Hyannis for Christmas the eight-hundred-mile trip would take between six and seven hours with one stop to refuel. That was the beginning of an extended dialogue about whether he would get enough leave to spend the holidays at home. "That's one thing I don't like about the Army. You are always in the dark until the last minute about things." Ty did make a quick trip home for Christmas—in his cousin's car, not an Army plane—and was back at Randolph by December 27. The next day, five cadets in his class washed out, in addition to two others who had done so earlier.

As he had been in California, Ty was eager to explore his new surroundings. He often spent weekends in San Antonio, sometimes joined by his cousin, Chan Cobb, the son of his mother's only brother, who lived in Houston and soon joined the Navy. Bobby, Chan's brother, who also lived in Houston and later joined the Signal Corps, occasionally tagged along. Ty and other pilots met at the Cadet Club at the Gunter Hotel, a large room staffed by volunteers where young service members could hang out, drink soda or beer, and attend tea dances. Ty and some pals spent one weekend touring Laredo, Texas, and went across the border to Nuevo Laredo, Mexico, where they visited markets, ate cheese enchiladas, and hired a carriage for a tour of the town.[32]

New Year's Eve, as he so often confessed about evenings out, "really cost me." He had a date with a "nice girl" spoiled by her engagement to another man, an officer who earned more than Ty did. "You just don't realize what you have to do in an outfit like this. I know you can't see my side of it." In a letter written near the time of his big night out, his traditional signature changed to "your loving son and broke boy." He was glad he didn't have a car to worry about, although that sentiment soon changed. As he began to make the case for buying a car, he surveyed the thirty-nine cadets from Oxnard who were in his class at Randolph. Thirty-seven had either bought a new or used car, had a car from home, had a car at home, or planned to buy a car as soon as they graduated. One wasn't sure he wanted a car. Then there was Ty, with a large question mark in a circle after his name. "I don't want one now but I was just giving you a picture," he wrote.

By January 2, 1941, he had passed his twenty-hour flight check, and his class finally had some underclassmen to tutor in drilling and table manners. Lessons in night flying began January 8, and Ty soloed at night at Randolph the next day. "It was about as much of a thrill as when I first soloed," he wrote, adding, "my instructor says I am a jitterbug. I can't seem to settle down. I don't know what the trouble is, but I am afraid of him."

Ty's class, 41-C, included pilots who had trained at several private fields before moving to Randolph and other Air Corps bases. Of his class of 391 cadets at Randolph, 344 remained, and Ty expected another 50 to wash out. "I hope I am not one of them." Mostly, he was confident. He was executing old skills in new airplanes that had different instruments and more of them. Soon, he practiced a spin using only instruments, "and it is really a job." Then he passed the forty-hour mark in solo flying. In the next breath, he complained that his two-year-old civilian dress suit fit him like "crepe paper that has been wet" and was shiny and uncomfortable. He estimated that replacing the suit and getting new shirts would cost $60.

In mid-January, about twenty minutes after fifty cadets took off, a hard rainstorm hit. In the 1940s, many airfields were grass, or in the case of a storm, mud. The pilots were all ordered to land at the closest air field, which for a few was

Randolph. Some of the planes landed at municipal airports from San Antonio to San Marcos forty-five miles away. Others were mired in farm fields. None of the pilots was injured, but it took several dry days before all the planes could take off and return to Randolph. Given that everyone was safe, Ty said he wished he had been flying that day.

A few days later, Ty was practicing a forced landing with his instructor. Instead of changing the propeller pitch to simulate engine failure, he accidentally cut the engine off at one hundred feet off the ground, making it a real forced landing. He flew under some wires and over a fence before getting stuck in the mud in a farm field. Farmers gathered around to get a look at the plane and talk to Ty and his instructor. Other cadets flew over to observe. The base eventually sent a car to pick them up. "It makes you feel awfully silly," he wrote. But there were advantages—one, knowing he could make a forced landing, and two, getting to know his instructor better.

As much as he came to enjoy aerobatics in the small plane at Oxnard, he said when he did them in the larger plane at Randolph it felt like blood was gushing through his body, making his head ache and his stomach flip-flop. At five thousand feet he was flying in snow, which evaporated before hitting the ground.

Although Ty was doing well in the air, he told his parents, he wasn't as successful in code class, in which he needed to decode ten words per minute. He made it to ten a few days later.

Despite a few setbacks, one thing is certain: Ty loved to fly. At his low points, he bucked himself up by saying that if he washed out of pilot training he could probably stay in the Army as a navigator. Sometimes he was goofy with joy. Other times, he was poetic. One evening in January 1941 at Randolph Field, Ty watched fellow cadets practice night flying. He described the scene in a letter to his parents in a rush of writing. Although he was observing others that night, the rush was his own:

Here I am sitting on the curb beneath the flood lights observing night flying. ... It is really swell out here listening to the drone of the planes taking off and in the air. And then the whine of the prop as the clutch engages with the engine. The warm up and then they are off after a signal from the tower. It is rather crisp out here and you can hear everything that goes on. A man with a sandbag is wanted here, the radioman is wanted in another place. A plane comes into the line and 4 or 5 mechanics are on it before it stops rolling. One cleans off the windshield, another is checking the instruments, a third checking the outside controller, and so on. Everything has to be perfect. The next pilot steps in and with a few last words from his instructor he gives it the gun taxies out onto the field, gets the go ahead signal from the tower and pushes the throttle full forward and there is a spurt and then a roar and the plane bolts ahead down the field as the motor picks up rpm's his tail leaves the ground then his wheels and he is in the air. He cuts the gun back and takes on a steeper glide up a hundred then two hundred feet. The turning props really sparkle in the flood lights. Then a turn to the left always climbing up to his zone of 1,300 ft. where he levels off and circles until the radio blasts out 'zone three from the tower come in for a landing.' He fumbles for his microphone putting his plane in a power glide at the same time heading for the last leg. He answers nervously into the mouthpiece 'Zone three Wilco' (for will comply with orders). He comes down three, five seven hundred feet at 130 miles an hour. He starts leveling off at 550 settling his airspeed down to 110, then he's coming near the field. He cuts the gun rolls the trim tab back, holding that plane at 500 feet. The altimeter now starts losing, the needle is going down. He establishes a 90 mile an hour glide.

He comes even with his landing leg. He banks the plane over half on its side and brings it around and levels it out. Then flaps, six turns of flaps a steeper glide brought to 80 miles an hour, he is coming close to the ground, the lights aren't far ahead. He starts breaking his glide gently and now more and more. His shadow rushes up to meet him, and he gets the stick way back in his lap, playing his feet on the rudder pedals to keep it straight. Then it settles down on the ground. The lights have flashed by and it is rolling to a stop. He eases on the brakes, swinging it around and heading back for the line, rolling up his flaps and settling his trim stop. Or else he gives it the gun and goes around again. From six in the evening until midnight this goes on.

He ended the letter with hopes his cousins or his parents could visit him soon or that he could visit Austin or see a few new movies. He also complained his parents' letters weren't arriving regularly.

Days later, Ty was excited to learn that his parents were planning to visit. He advised them to have a mechanic check the car before they left Hyannis and before the return trip. He suggested they could overnight in Muskogee, Oklahoma, to see a friend, and on the way back, they should stop in Ponca City, Oklahoma, "one of the most beautiful towns in the U.S." They should arrive on a Wednesday, he said, because he would be free from 4 p.m. to 7:30 p.m. While he was busy Thursday and Friday, they could take a side trip, perhaps to Houston, where Chan and Bobby Cobb lived, or to the beach at Galveston, adding, "But it is your trip." He would be free Friday for supper and from noon to 7:30 p.m. on Saturday. "I think I can arrange things so that you can inspect the airplanes, hangars, etc.," and he would get passes so they could enter and leave the base without an escort. "That way you can see us fly both at day and night."

Ty's letters home resumed three weeks later, on February 8, when he wrote he was nearly done at Randolph. He completed his flying course the day Hans and Shy left, and he told them he followed their car in his plane, wagging his wings. They apparently didn't notice. His options for advanced flight training were bombardment, observation, or pursuit. "I don't know which branch I want either, but they do most of the deciding," he wrote, referring to the Army. Ty ruled out pursuit training, because the smaller, high-speed, and highly maneuverable planes threw the pilot all over the place. He also concluded he was too tall for the smaller planes. He and his friend from Kearney, Jim DeWolf, were assigned to report to Kelly Field on February 10 with fourteen of their classmates from Oxnard, where they would train to fly bombers. "Kelly has a name," Ty wrote, adding that Uncle Thomas E. Kokjer had trained there during World War I.

He met some Kelly cadets one weekend at the Gunter, and they told him the only way pilots there washed out was in a coffin or because of the physical exam, including blood pressure—the only part that worried Ty. Pilots didn't get much instruction at Kelly, the cadets said. They were there to learn to fly new planes and to practice flying in formation. Generally, Ty found Kelly to be just as swell as his other assignments, with the exception of having to get up fifteen minutes earlier. But the barracks were new. He wrote that they were treated as if they were already officers. The commanding officer spoke to them. They were told about government insurance, were fingerprinted, took physicals, and applied for their commissions as officers.[33]

Next up was training in a twin-engine plane, learning to fly in formation, and plotting and completing long-distance flights. The cadets learned about the effects of high altitude and using oxygen. Ty was told to eat less but to add foods that would aid his eyes for night flying: eggs, lots of butter, green and yellow vegetables, cheese, and milk. They were to fly the BC-1 (basic combat), also known as the AT-6, which had more gadgets than the planes at Randolph as well as a mount for a machine gun. It was also his first plane with retractable landing gear. He practiced on the ground in a Link Trainer, which tilted and rotated to simulate flight, but his first flights at Kelly were in an AT-6A, an advanced, single-engine plane. He finally flew in the BC-1 with his instructor February 23.[34]

Knowing the Army would make the final decision, Ty and DeWolf listed flying bombers out of Honolulu as their first choice for postgraduation assignments. Alaska was second, and the Philippines were a distant third. They could fly transport planes, ferrying freight and soldiers around the country. The worst, the pair decided, would be assignments as flight instructors in the United States. In the end, Ty wrote, it didn't really make any difference. He expected to graduate in mid-April.

Around this time he wrote to Mary Tice, hoping to see her before he deployed. "I sure would like to see her, if nothing more than to satisfy myself that she is as swell as ever or else so I can forget her altogether. I think about her a lot," he wrote to his parents. Mary, who was about to start her senior year in high school when she and Ty met the summer after he graduated from high school, replied with the news that she had quit her job at the drugstore and moved from Gordon, Nebraska, near Hyannis, to live with her sister in Indianapolis, Indiana, to attend business school. Ty decided that he and Gramp needed to take a road trip to Indiana on Ty's predeployment leave.

Meanwhile, flight instructors sent their students up to fly in bad weather and low ceilings. "We really had fun this morning. We went high above the clouds—it was dark and all gloomy down below but up above 7,500 feet it was clear as a bell, and the sun shining down on the clouds making them look white and fluffy." That's a view thousands of people now take for granted every day, but in 1941, flying high above the clouds must have seemed like a rare adventure.

In early March, Ty began to practice formation flights, which he came to regard as another adventure after first not liking them. His instructor was in the lead plane, and Ty and another cadet followed, one on each wing. "He would dive in through the clouds and come whooping up and make steep turns. It's just like a game of follow the leader, and on those turns, it is much more like crack the whip. Or when I'm on the outside of a turn, I would shoot way over by a half mile or so and have to come back and catch up again." Later formations included nine planes, which were presumably much less adventurous. On a solo flight the morning of March 10, he got lost on purpose, turned on the radio beam and compass, and found his way back.[35]

Ty was almost done with ground school, and after dropping to 174 pounds, he passed his physical. He was studying the officer's guide and hoping for Hawaii. "The Philippines don't look too hot." When the class above his graduated, Ty had dinner with Hiram Messmore, a Lincoln acquaintance, and Hiram's parents. Hiram's father, Frederick Messmore, was a Nebraska Supreme Court judge and a friend of Ty's Uncle Emerson Kokjer, who was deputy attorney general for the state from 1939 to 1947. Hiram and fourteen other pilots from his class were bound for Manila.

As Ty was thinking about his posttraining assignments, Joseph Grew, the U.S. ambassador to Japan, told his superiors that he had heard rumors from several sources that "seemed fantastic." Japan planned to attack Pearl Harbor in case of trouble with the United States. His earlier warnings that war with Japan "may come with dangerous and dramatic suddenness" were "as usual ignored" by the State Department, where officials thought Grew was "old fashioned and honorable but gullible," and that Japan was bluffing, according to the historian John Toland.[36]

"A war that need not have been fought was about to be fought because of mutual misunderstanding, language difficulties, and mistranslations ... irrationality, honor, pride and fear—and American racial prejudice, distrust, ignorance of the Orient, rigidity, self-righteousness, honor, national pride and fear," Toland wrote.[37]

In February, German soldiers had begun rounding up Jewish families in Poland, and over several months in the spring, Japan seized control of rice and rubber in Indochina (now Cambodia, Laos, and Vietnam), Thailand, and the Dutch East Indies (now Indonesia). Shortly after those actions, a cabinet controlled by conservatives and military officials took office in Tokyo, replacing more moderate men.[38]

The United States sent two thousand Army troops to the Philippines on April 22. The next day, the America First Committee, which supported defending the United States rather than entering foreign wars, held its first public rally, with the pioneer pilot Charles Lindbergh as a featured speaker. The isolationist group and Lindbergh drew rebukes for antisemitic statements.[39]

Whether Ty was oblivious or realistic about his future in the Army and a looming war, he was having so much fun, he feared it wouldn't last. "DeWolf and I are all the time talking and both think something will happen as it seems so good." They were planning a postgraduation "gala dream trip home," with stops in Kearney, Hyannis, and Lincoln.

The friends' foreboding became reality on Friday, March 21. DeWolf was driving his car back to Kelly from San Antonio on a dark road and missed a sharp turn. The car ran off the road, with Ty in the passenger seat, and smashed into a tree in the side yard of a house. Ty's right leg was badly broken about three inches above the knee. He also broke two or three ribs and lost two teeth. DeWolf's injuries were less severe. He would fly again in three weeks, graduate on time, and deploy to Hawaii. For Ty, it would be four months if his leg healed properly, pushing graduation from April to August. If healing was less than optimal, combat would be out, but he could stay in the Army as a flight instructor. DeWolf felt guilty and assumed his parents were mad at him. "He feels so bad he doesn't even want to go home." The car was a total wreck. Ty said the homeowners were surprised that they lived.

Both became patients at Fort Sam Houston Hospital midway between Randolph and Kelly fields. Fellow cadets came to visit. The room had windows that were big enough to allow Ty to see what he was missing out on—planes flying by. He hurt all over. "Then of course I hate the bed pan." He wouldn't get a cast until his chest was better and his temperature came down. In the meantime, he was flat on his back with his leg immobilized. "You should see how they have me strung up in this bed ... ropes, pulleys, weight bags, iron, tape ... everything." One night he "got all tangled up in a bed pan," throwing off the suspension system for his leg. All that slowed his letter writing. He sent his parents a sketch of himself with his leg strung up and a diagram of the accident. "I wish I had sleeping sickness at the same time and could wake up when everything was OK again."

By April 2, he was in a cast after three hours of surgery. Instead of plates to stabilize his leg, the surgeons used "a new 20th century technique, pins." He would be in the cast for twelve weeks. The next day, he started to walk, meaning he could dispense with the despised bedpan. Chan came to visit, and Ty had many get-well cards. Groups of doctors checked on his progress and the modern cast. DeWolf bought him a shirt and a pair of pants. A nurse ripped out part of a seam in the right leg so it would fit over his cast, and he wore regular clothing for the first time on April 12. His doctors said he could go out. Five days later, a friend took him to the hangar at Kelly Field, where he met with his instructor, who told him he might have to repeat the entire ten-week program. On the way back, Ty and the friend stopped for a malted milk at the Pig Stand, a drive-up, drive-through, pig-themed restaurant in San Antonio that is still in business.[40]

Ty passed the next few weeks reading, writing letters, and visiting with friends who stopped by. Radio was piped into the ward room, and Ty said when his parents were listening to the news at noon, they should know he was too. He listened to Fred Waring's band, boogie-woogie, or jitterbug. Zane Grey's Western adventure novels were his favorite books. One evening helped break the quiet spell. He rented a car, and Miss Hutton, a nurse, drove. They met some of Ty's friends at a restaurant, and Miss Hutton found other nurses to be their dates. They ate dinner, and everyone but Ty danced until 2:45 a.m. He later assured his parents that his outing was not the reason the hospital was enforcing a midnight curfew. The rule had always been there, he said, but no one had paid any attention to it.

He was often bored, mulling over plans for the near and more distant future. He began to write to his parents about buying a car, as it was "imperative" for officers to have cars, and prices would certainly surge in 1942. He expected he would be well enough to drive by July. During his time off, he could drive to Indiana to see Mary Tice, who continued to write to him.

Things picked up on April 26, when Ty achieved a milestone. He took a bus to Kelly on his own, ate dinner, and rode back. But it happened to be graduation day for another class, and he saw cadets celebrating with their families. "I think this is the hardest thing I have ever had to take in my life in the way of wanting something. Just think, I would be all done now and have my commission and I would be coming home at the end of next week for a short stay before going to my next station."

DeWolf and Ty's other former 41-C classmates would graduate at the end of May.

The following week, some of his friends took him out, then made fun of him for drinking orange juice instead of beer. In early May, the hospital dentist cut out his broken teeth and took an impression for a partial plate. By now most of his friends had graduated, and life was dull. He found entertainment where he could. An accordion concert by local kids "was a lot of fun." Ty was flattered when his cousin Margaret Cobb, Chan and Bobby's younger sister, wrote to ask Ty to teach her how to twirl a baton when he got home. In mid-May, Hans drove to San Antonio and took him home for a month of leave. Ty's ability to overcome boredom and enjoy things and people around him would prove to be a valuable trait in the months to come.

The Hotel Hyannis opened as the DeFair in 1898. This
photo was taken shortly after the Kokjers began managing
the hotel in 1939. Family photo

The Hotel Hyannis—sometimes called the Hyannis Hotel—looked nearly the same in 2017 as it did in the 1930s. The railroad tracks are just beyond the car parked across the street from the hotel on the right. Photo by Jody Beck

Hans, Shy, and Ty Kokjer outside the Hotel Hyannis. They took over management of the hotel in 1939, and Ty was last there in October 1941. He wrote about returning to the hotel after the war, but also wrote about living elsewhere. Highway 2 and the railroad tracks are to the left. Family photo

Chapter Four

Journey to a Distant Future

*"The sparkle went out of Manila in the spring of 1941.
War was coming and we all knew it."*

—General Jonathan M. Wainwright, commander of the
North Luzon Force, Philippines

While Ty was on leave because of his broken leg, Hans and Shy closed the hotel for a weekend in early June and hosted a family reunion. Uncle Tom Kokjer posted a sign: "No paying guests wanted." Among those at the reunion were Hans's older sister, Meta Key, and her daughter, Dorothy; his youngest brother, Emerson; and Emerson's family—his wife, Winifred, and their children, Phyllis, 15, and Janet, 10. Phyllis was grateful that Ty, then 21, treated her as an equal and brought her along when he visited friends. Janet remembers playing with their younger cousins and getting ice cream from the hotel kitchen whenever they wanted it. They all spent a day at the Abbott family's ranch. "Dad looks great in a 10-gallon hat. Uncle Em did a great job rustling. Aunt Meta, Aunt Win, Dorothy, Phyllis and Janet did a fine job looking on. I had a great time on crutches," Ty wrote later in his diary. One photo shows Phyllis and Janet on horseback, with an Abbott family friend.

Ty started back to San Antonio early on June 12, 1941, as a passenger in his new car with friends along for all or part of the ride. After doctors removed the pins and his cast on June 18, he still needed crutches for a few days. On a date shortly after he stopped using the crutches, he tried to dance for the first time, "and it worked out so well, we danced about a half a dozen dances." He drove to Galveston for a weekend at the beach with a friend and his cousins, in part to swim, as his doctor ordered, to help his leg recover. On June 23, he headed back to Nebraska to continue his recovery. By early July, he was able to play golf with his dad and relatives in Sidney, and he drove to Funk to visit his grandfather, U.S. Cobb, who operated the grain elevator there. He saw Marvin Metzger, his art student friend, who was at home in Gordon, eighty miles north of Hyannis.

Ty returned to Texas on July 24, passed his physical, and moved into his old barracks at Kelly Field with the cadets in his new class, 41-G. At first, Ty expected to finish training in mid-August, but he amended that to September. He wrote to tell his parents how much he had enjoyed his time at home. He experienced his first Texas summer: "It's really hot down here." At the time of the accident, he had thirty hours of solo flying. He needed fifty-nine. After flying on July 29 for the first time since March, he wrote, "It seems now like I have never been away from it, although I did make some mistakes today." He and the other cadets prepared for an inspection visit by General Henry A. "Hap" Arnold, the commander of the Army Air Forces, as the Army Air Corps had recently been renamed. Ty didn't comment about meeting one of the country's star aviators, but noted, "We have to be on our best behavior."

Arnold learned to fly from the Wright brothers, and his career tracked the development of what became the U.S. Air

Force in 1947. When World War I began in Europe, the United States had six military airplanes and fourteen pilots. The first American pilots to fly in combat flew French planes. By 1926, the U.S. Army Air Corps had 916 officers, 8,725 enlisted men, and fewer than 1,000 planes. The numbers rose quickly after 1938, when Arnold became chief of the Army's air division. By 1942, the U.S. Army Air Forces included 26,500 men and 2,200 planes. Three years later, there were 2.25 million men and women in the Air Forces, which had nearly 64,000 planes.[41]

In early August, Ty was planning the first of several cross-country flights, point-to-point-to-point flights that could take several hours or most of a day or night. On his first, he flew north to Fredericksburg, northeast to Temple, back south to Austin, and west to Kelly. He flew at eight thousand feet above "big, fleecy cumulus clouds." He could see the plane's shadow on the clouds and flew in circles around the shadow, sometimes glimpsing the ground in gaps in the clouds and at other times, rainbows. Another day, he flew at 22,800 feet, high enough to require oxygen. The plane felt different at that height, he wrote, without as much air to support it. He reported that the gas in his stomach expanded to three times normal, and he ended up with a headache and ringing ears.

In addition to the view of clouds and shadows that week, Ty could see the end of his training. He signed papers, once again listing Hawaii, Alaska, and the Philippines as his preferred assignments. The idea of war hovered in the background. Ty and his family read newspapers and listened to daily radio newscasts. In hundreds of letters from 1937 to the moment Ty received orders to deploy to Manila, he rarely wrote about the risks he faced or that he would fight in a war. Were they, like most Americans, "blissfully unaware" of the possibility of war, despite new government agencies to manage a war and industries ramping up to make ships, planes, jeeps, and walkie-talkies, as author Winston Groom wrote? (Groom, who alternated between writing fiction and nonfiction, is best known for his novel *Forrest Gump*, which became a hit movie.)[42]

In response to what Japan saw as a stiffening of U.S. conditions for peace, the island nation made a preliminary decision July 2, 1941, to move south to consolidate its grip on Indochina, Siam (Thailand), the Burma Road, and the Dutch East Indies. The United States had some information about Japan's plans because it was reading its diplomatic cables, but had not broken its military codes. The Philippines, but not Pearl Harbor, were thought to be a likely target for attack.[43]

The islands' military strength looked good on paper. But Douglas MacArthur's plan to defend the entire Philippines coastline, which is longer than that of the United States, with fifty PT boats was unrealistic, even more so when he got only nine.[44]

After thirty-three years in the Army, including five years as chief of staff, MacArthur retired and in 1936 became a civilian military adviser to the Philippine government. It was a familiar place for him. His father, Brigadier General Arthur MacArthur Jr., led the first forces in the battle to secure the U.S. hold on the Philippines, in August 1898. Shortly after Douglas graduated from the U.S. Military Academy in 1903, he was posted to the Philippines as a second lieutenant with the Third Engineer Battalion. He returned for a year in 1922 after four years as superintendent of West Point. On July 26, 1941, President Franklin Roosevelt made war likely when he froze Japan's assets in the United States and cut its access to U.S. oil. Japan imported almost all of its oil, and 80 percent of that came from the United States. Roosevelt returned MacArthur to active duty in the Army the same day.

Two days earlier, Ty had returned to Kelly, and he played cards with new classmates the night of the 26th. He didn't know about Roosevelt's decision and how it would affect him. Japan had set midnight November 30 as the deadline to conclude negotiations with the United States. Although negotiations continued as both sides needed time to prepare, the United States rejected Japan's last two peace offers on October 16 and November 26, knowing the final rejection made war more likely. On November 5, Japan's soldiers and sailors began their final preparations for the assault on Pearl Harbor. It would attack the U.S. base in Hawaii in December because there would be a full moon, and winter weather would preclude a later attack there. Japanese commanders knew the U.S. fleet usually left port Monday morning and returned Friday afternoon.[45]

In John Hersey's first book, *Men on Bataan*, published in the early spring of 1942, he created a portrait of MacArthur seven months before the United States was drawn into the war. MacArthur was convinced that Germany had told Japan "not to stir up any more trouble in the Pacific." Like many others, MacArthur described Japanese troops as inferior soldiers fighting with inferior weapons. MacArthur assumed Japan wouldn't attack the Philippines before April 1942, when he would have enough soldiers, pilots, and arms to win easily. But he claimed the United States, Great Britain, and the Dutch could defeat Japan with about half the forces already deployed in the Pacific.[46]

General Jonathan M. Wainwright was the Luzon Island commander and reported directly to MacArthur. "The sparkle went out of Manila in the spring of 1941. War was coming and we all knew it," he wrote in his memoir. Most U.S. dependents were evacuated in May, and soldiers began drills with an eye on war the following spring. Wainwright noted how ill-prepared everything was. A history of the war in the Philippines echoed Wainwright's assessment. Anti-tank and artillery divisions were not at full strength or fully armed. Professional soldiers in the Philippine Scouts were well trained and effective, but new recruits had little training.[47]

As preparations continued 8,400 miles away, Ty was finishing his training. He made cross-country flights during the day and at night, hoped he could buy a used Army trunk instead of a new one, discussed whether it made sense for his parents to attend his graduation, and wondered if they had proper clothing that fit—both Hans and Shy struggled with their weight. On a Sunday in early September, he saw the movie *Dr. Kildare's Wedding Day*, with multiple romantic and medical plot twists. In the newsreel that preceded the film, he saw his friend Wally Churchill in the Philippines. "He sure hated to go over there."

Ty was still thinking about Mary and marriage and what he thought it took to have a successful one. Two people interested in the same things, a home, and children, he wrote to his parents. He figured they were laughing at his musings about a future domestic life. He asked if they would consider bringing Mary to Texas for his graduation. Later, he wondered if visiting Mary before he headed overseas could make things worse, especially because it would be at least a year before he could marry. While he was in the Philippines, he wrote that he would want to start educating his children before they entered school with "instructive games and toys," plus trips to places they would learn about in school.

On September 14, Ty got the news he had worried over. He was going to the Philippines. He wrote the news to his parents that day: "I am going to the Philippines with 11 other fellows. ... 33 are going to Hawaii—I guess I lost out there." But he added, "I intend to enjoy myself before I leave." Friends from class 41-G, including Leif R. Kloster and Robert G. Bjoring, also drew Manila. All three were assigned to the 27th Bombardment Group to fly light bombers. Sixty-three cadets from Ty's original class, 41-C, were already there. A few others were assigned to Panama and Alaska. He reasoned that it would be more expensive to live in Honolulu.[48]

"Well today I finished my flight training at Kelly Field," Ty wrote September 5, 1941. He was hoping for some leave but was turned down because of all the time he had off while his leg healed. He would have to remain in Texas, with nothing official to do, and would have to find and pay for his own housing until graduation on September 26. There were parties, however, including a picnic September 13 with athletic competitions. He did not do well at croquet. Two days later, his instructor hosted a cocktail party. Ty wasn't excited about military social customs. "I don't think I will ever become a drinker because I don't like it." The party turned out better than expected, however. After the first drink, everyone mixed his own. "So I remembered to forget to put any of the liquor in and I don't think anyone was the wiser. ... I enjoyed being with the fellows and talking over our flying futures."

And still, the war seemed abstract to Ty and his family. Most Americans opposed getting involved in what they perceived as other people's fights. In a poll published October 4, 1940, 83 percent of Americans opposed sending troops overseas. In his September 14 letter, Ty made one of his rare comments about war and danger. "The foreign situation doesn't worry me in the least. And don't you worry about it either. I might have been sent to Iceland and then you would have had a cause for worry." The United States had taken over the defense of Iceland earlier in the year, and would later do the same in Greenland, once Denmark fell, to prevent Germany from establishing bases in the Atlantic Ocean. Such bases would have created a threat to Great Britain and North America. Although the Philippines would be twice as far as Iceland, the distance from home didn't bother Ty much, unless he or one of them got sick. "I know I will miss you terribly and I think you will miss me a little." Two years, he said, wasn't such a long time.[49]

On September 19, Ty turned in his flight gear and parachute, attended more parties, and began sending out graduation announcements. All the celebrations were expensive, but worth it, he decided. "I have a wonderful education behind me. Best of its kind in the world. If I were out in civil life right now, I could really cash in." In a letter to Ty that came back after the war started, his grandfather U.S. Cobb said he enjoyed the graduation trip to Texas, when he ate chili and goat chops for the first time, noting he preferred old-fashioned American food.

After being told he wasn't eligible for a furlough, Ty got a week off. Curiously, there's no indication he used that time to go to Indianapolis to visit Mary Tice. After a few more days back in San Antonio, he drove to Hyannis, arriving on October 4. He and his parents drove Ty's car 140 miles to Tom Kokjer's home in Sidney that day and left for California the next morning, arriving at Uncle Roy Kokjer's Larkspur home October 7.

Two days later, Ty checked in at Fort Mason, now part of the Golden Gate Recreation Area. For many years, it was a stepping-off point for soldiers and military cargo, including two-thirds of all troops sent to the Pacific. Ty's orders hadn't arrived. Shy later wrote in her scrapbook: "We had a continuous week of pleasure with you and our family there." His orders came through October 15, the day his parents caught a train back to Nebraska. Ty missed the ferry after leaving his parents at the train station, meaning they could have spent a final few minutes together. "We wish we had known, we might have been with you. These min. were precious," Shy wrote. "Our parting wish was that we would see the same dear boy two years later. May that wish come true."[50]

Ty was to sail October 26, 1941, on the USS *Hugh L. Scott*, a passenger cruise ship. Bob Bjoring was to board the *President Harrison*. Over the next few days, Ty enjoyed outings with relatives and friends, including Bjoring. They went to the races, visited Fisherman's Wharf, and arranged dates with some Army nurses for a gala dinner at an expensive restaurant. Ty and another pal and their dates slipped out before the bill came. "The next morning at breakfast Woody and Kokjer were having a hilarious time at their joke on us," Bjoring wrote in his memoir. Before Bjoring's check for the entire dinner could bounce, Ty and the other officer wired money to Bjoring's bank for their shares of the dinner. Shortly thereafter, Ty put his car on a ship to Manila and visited Hamilton Field, in Marin County, which was preparing to send twenty-five B-17s to the Philippines. Ty guessed that was what he would fly and wished he could fly one of the planes there himself to save time. The four-propeller Boeing, known as the Flying Fortress, and the B-25 Mitchel Bomber were the two long-distance, workhorse bombers of World War II.[51]

A week later, Ty dashed off a quick note to his parents and boarded the ship. For Ty and other officers, the three-week trip to Manila was a luxury cruise, with stops in Honolulu and Guam. He described his accommodations and the food in a letter mailed November 2 from Honolulu. "Well here I am on the high seas. ... This trip is really swell." They had "roast beef like home. There are olives, salted nuts, celery, salads. Everything you could think of, served well." Each table seated seven officers, who wore jackets and sat in armchairs. They dined on china, used silver creamers and sugar bowls, and were served by a Filipino waiter. One member of the 27th Bomb Group described the food as coming from "the kitchen of Henry VIII." Ty's shared stateroom had two beds and running hot and cold water. In the ship's lounge were comfortable chairs, phonograph records, and a piano, although Ty said it was out of tune. There was deck tennis and Ping-Pong as well. Neither Ty nor Bjoring were among those who were seasick. "Every time the boat rocked a little extra hard, they got the funniest look on their faces. It really tickled me even though I was in sympathy

with them," Ty wrote. "I only wish you could both be along."
Later in life, Hans and Shy became cruise enthusiasts, and
based on Ty's description of his own trip, he likely would have
joined them. Surely he had to push the thoughts of this cruise
from his mind the next time he boarded a ship in late 1944,
which took him from the Philippines to Japan.[52]

Other officers, older men of higher rank and more experience, enjoyed their time at sea, but they were more realistic
about what lay ahead. David L. Hardee had fought in World
War I, served in the Philippines from 1929 to 1932, and was a
lieutenant colonel when he sailed for the Philippines a second
time. "Coming events cast their shadows and all felt that we
were going into war and many aboard would never see their
native land and loved ones again. None of us felt that we
would have a good time for two years at the very pleasant task
of training the Philippine Army." Edward Dyess was a captain
who led a pursuit squadron when his ship sailed for Manila on
November 1, 1941. He assumed he was heading to war, based
on news reports about increasing tension with Japan, although
many of his friends were doubtful. "We were to remember their
laughter in the weeks that followed. It reflected the attitude of
the nation. It was an explanation, in a measure, of America's
bewilderment and shock at the bombing of Pearl Harbor. Yet
we knew nothing the civilians didn't know."[53]

Ty watched from the deck as the harbor pilot picked up the
Scott early on the morning of November 2 to guide the ship
into Honolulu. "It's a beautiful looking island from out here."
During the ship's stay in port, Ty was made assistant officer
for the enlisted men's mess, a far different food service than
the one enjoyed by the officers. He described the mess rush
"about like fair time at home," when people from all over came
to the Grant County fair in Hyannis, with many of them staying and eating at the hotel. On the ship, they had to feed one
thousand enlisted men in a cafeteria line three times a day.

The *Scott* remained in Hawaii for a few days to await other troop transports. Ty used the time to visit Pete Holden, Shy's uncle, who made a living renting out holiday cottages and writing occasional pieces for the local newspaper. Ty also found classmates from 41-C and 41-G who were stationed there. One took him up for an aerial tour of the island and let Ty pilot the plane for thirty minutes.

From Honolulu on, the troop transports would be accompanied by a military convoy, he wrote his parents. "Now don't worry as nothing will happen. They are just taking every precaution."

Ty sent another letter from Honolulu and one from Guam. Hans and Shy both replied to the Guam letter. She was pleased that he sounded happy, that he was able to fly while he was in Honolulu, and that he spent time with Uncle Pete. They told Ty they had ordered a subscription to *The Grant County Tribune* for him. The paper, Shy wrote, circulated in twenty states, but Ty would be its longest-distance subscriber. She estimated it would take the paper about a month to reach him.

Ty's ship arrived in Manila on November 20, 1941. He sent his parents a telegram confirming his arrival.

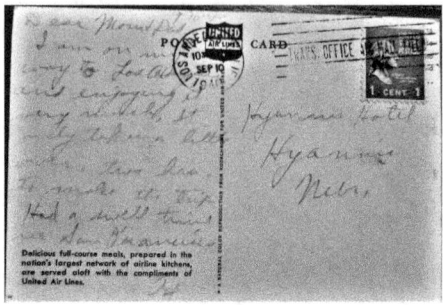

After visiting his uncle's family near
San Francisco, Ty flew to Los Angeles
on United Air Lines (later United
Airlines), to begin his pilot training in
September 1940. Photo by Joseph W.
Brown

Ty praised the fine dining, with table
cloths and chinaware, and the two-hour
flight. Photo by Joseph W. Brown

Ty learned to fly in the open-cockpit Stearman PT-13 Kaydet biplane at an airfield leased by the Army in Oxnard, California. Stearman, which was acquired by Boeing, made thousands of training planes. Ty and is instructor could pilot from either seat. Ty's first "swell" solo flight was in the Kaydet on October 1, 1940. Family photo

Chapter Five

War Erupts in a Tropical Paradise

*"JAPANESE FUTURE ACTION UNPREDICTABLE BUT HOS-
TILE ACTION POSSIBLE AT ANY MOMENT."*

—Message from Secretary of War Henry L. Stimson to
General Douglas MacArthur, November 28, 1941

As Ty arrived in Manila on Thanksgiving Day, he wasn't thinking about war or politics. He and many other officers expected two years of adventure in an exotic city eight thousand miles from home. Even though it was Ty's third choice, for most military personnel, Manila was a desirable assignment.

The city of 684,000 residents on Luzon Island was called "the Pearl of the Orient." Other superlatives abounded. Author Judith Pearson called it "a blend of cosmopolitan metropolis and country life rooted in centuries past." MacArthur and his family lived in a penthouse suite at the Manila Hotel, one of the few "air cooled" buildings in the city. From his balcony, the general could see thirty miles across Manila Bay to the mountainous Bataan Peninsula, which protected the city's busy harbor, the best deepwater port in the southwest Pacific.[54]

One author described Pier 7, where Ty's ship docked, as smelling of "fish, exhaust fumes and horses." Bob Bjoring described "a pervasive sweet/sour smell with a smoky flavor."

He reported that all of his 41-G classmates, including Ty, were taken to a Thanksgiving dinner, put up at a hotel, then given the weekend off. They began to explore Manila. On Monday morning, they were taken to Nichols Field, the air headquarters south of Manila, to check in and begin their orientation. They were told to order dress uniforms.[55]

The salaries of low-ranking officers "supported a lifestyle in Manila that only the ultrawealthy could afford back in the States, where the Great Depression still held sway. Among the usual perks were a fashionable apartment, a full staff of servants, access to exclusive golf, tennis, polo, and drinking clubs —and plenty of leisure time to enjoy all this," author Bill Sloan wrote. As a second lieutenant, Ty didn't live that luxuriously, but he and his three roommates at a downtown hotel employed a house boy and a laundry boy. He bought a golf outfit and played every day. He bought Christmas presents for his parents and cousins, which made it to Hyannis in early January. Before his car arrived, he discovered he probably didn't need it. Rides in cabs or calesa pony carts cost only a few cents.[56]

The city was dotted with churches, libraries, schools, universities, botanical gardens with zoos, commercial warehouses, department stores, shops, private clubs, air-conditioned movie theaters, bowling alleys, swimming pools, tennis courts, and nightclubs. The city's residents bought snacks from street vendors, haggled for clothing and hats in crowded bazaars, and spent hot nights near the cool waterfront. After three hundred years of Spanish rule, followed by forty years as an American possession, Filipinos spoke Spanish, English, Tagalog, and dozens of other Filipino dialects. Nearly everyone belonged to the Catholic Church. Poor residents lived in bamboo huts away from the glamorous neighborhoods near the bay. As the United States took control of the country in 1902 after the Spanish-American War, its main exports were sugar, tobacco, rope, and coconut oil. The self-governing commonwealth was scheduled to become independent in 1946, not fast enough for some Filipinos who resented U.S. control.[57]

Japan set midnight, November 30, as the deadline to reach a peace agreement with the United States. But the quest for peace had effectively ended before then. Months of failed negotiations led Japan to decide in November it would attack Pearl Harbor the following month. Secretary of War Henry L. Stimson got permission from Roosevelt on November 28 to warn MacArthur by radio of "only the barest possibilities" Japan would agree to further negotiations: "JAPANESE FUTURE ACTION UNPREDICTABLE BUT HOSTILE ACTION POSSIBLE AT ANY MOMENT PERIOD IF HOSTILITIES CANNOT, RE-PEAT CANNOT, BE AVOIDED THE UNITED STATES DESIRES THAT JAPAN COMMIT THE FIRST OVERT ACT." MacArthur was allowed to take actions to protect his forces. A similar message sent to the Hawaii commander told him not to alarm the public.

The same day, *Time* magazine's correspondent in Manila, Melville Jacoby, reported to his editors about increasingly dire predictions for the likelihood of war. On December 6, he wrote, "Informed sources this evening are seeing mounting chances of war," adding that Japanese planes had been spotted nearby. It turned out to have been a rehearsal.[58]

The December 8 issue of *Life* magazine (*Time*'s sister publication) featured a photograph of Lieutenant General Douglas MacArthur on the cover, wearing an Army shirt with three stars on the shoulder, gazing confidently into the distance. He got a fourth star later that month. Before the attacks on Pearl Harbor and Manila, Americans mostly ignored—or were ignorant about—the Philippines. Journalist Clare Boothe Luce interviewed MacArthur in October and wrote the story about him for the popular weekly picture and feature magazine, price 10 cents. She reported that budget cuts in previous years slowed MacArthur's progress in building up the Filipino armed forces, and perhaps foreshadowed defeat. "He damned the military myopia at home and abroad, and flatly predicted that, if it were not remedied, the Philippines must drop like an over-ripe plum into the Japanese basket."[59]

Air mail Clipper service ended just before the war started, and Hans and Shy's letters began to come back to them stamped "Return to sender/Service suspended." Ty later wrote that he kept the last two letters from his parents that he received in Manila. In the first returned letter, dated December 3, Shy asked about Ty's Thanksgiving dinner and told him about the Christmas trees Hyannis was putting up around town, complete with electric lights. The cowboys from the Sibbitt family ranch had been in for dinner on Sunday, when dinner was still a midday meal. She wanted to see Ty in his new dress uniform and had bought a white dress she could wear if they were able to visit him, perhaps in a year. She also wanted to talk to him on the phone but realized it would be expensive and decided they could wait until he had been there longer. "We don't worry about you, Ty. We know you are well taken care of and that you are enjoying the scenery. ... We are sure Japan will find herself before she gets into it with the U.S.A." Her feelings weren't unique. Most Americans underestimated the intelligence and skills of Japanese soldiers as "an enemy who could not see straight, could not shoot straight, could not keep his planes on course or drop his bombs on target." If war came, Ty's parents expected it would be short.[60]

Roosevelt made a final appeal to Emperor Hirohito for peace, but concluded the time for negotiations had ended. On December 4, the United States intercepted a coded message telling the Japanese ambassador to burn sensitive documents and destroy the code machine. An intelligence officer could see smoke from Massachusetts Avenue in front of the embassy.[61]

On December 6, Ty and members of his squadron played softball against the elite Manila Polo Club team. The polo club team won, but Ty didn't report the score. The evening of December 7, Ty attended a party for the 27th Bomb Group at the Manila Hotel. (Because of the International Dateline, it was still December 6 in the United States.) It was a gala affair, historian Louis Morton wrote, with "the best entertainment this side of Minsky's," a New York burlesque theater, to honor General Lewis H. Brereton. He was the new commander of the Far East Air Forces in the Philippines, as the air wing was called. Few at the banquet noticed when Brereton slipped out to take phone calls about increasingly dire warnings of a possible Japanese attack somewhere in the Pacific. Ty didn't say much about the banquet for eighty officers, but he liked the dessert: "swell ice cream—like homemade."[62]

For most in the United States, and in Washington, December 6 was a normal day, "nothing ominous in the atmosphere, no forebodings or shadows cast by coming events," Morton wrote. In the early afternoon on December 7, the Japanese ambassador called on Secretary of State Cordell Hull to say negotiations were over. It was past sunrise in Honolulu, and bombers had already struck the U.S. fleet. It is likely the ambassador did not yet know the attack had taken place, but an angry Hull did.[63]

At Griffith Stadium that afternoon, the Washington football team was playing the final game of the season against the Philadelphia Eagles when the announcer began to page generals, admirals, colonels, and captains to report to their offices. But it wasn't until the game ended, 20–14 in favor of Washington, and twenty-seven thousand joyous fans left the stadium, that their moods were crushed. They were met by newsboys hawking special editions of newspapers. War had come to the United States.[64]

"The world tilted; football lost all importance," S. L. Price wrote in *Sports Illustrated* in 1999. Americans who were relaxing at home, perhaps listening to the Giants–Dodgers football game on WOR or the New York Philharmonic on CBS Radio, heard about the attacks shortly before 2:30 p.m. when the broadcasts were interrupted. S. Lawrence Heisinger, a lawyer in the National Guard, had gone to the Philippines for what he thought would be a six-month assignment. His family heard the news on a car radio as they left a Sunday morning church service in Sacramento, California. "Our war, our very personal war, had begun," his son, Duane, then 10, wrote many years later. Military recruiting stations began the week with lines out the door, and government officials moved the Declaration of Independence, the Constitution, and other documents to Fort Knox. Some Washingtonians were so angry with Japan that they chopped down a few of the cherry blossom trees at the Tidal Basin. Japan had donated them during the Taft administration. Others gathered in front of the Embassy of Japan on Massachusetts Avenue Northwest.

That day, MacArthur told Wainwright and the North Luzon Force to be ready to move. "The tension could be cut with a knife," Wainwright wrote in his memoir. "I got a good night's sleep. It was the last decent night's sleep I was to have for three years and eight months." When word of the attack on Pearl Harbor came in the early hours of the next morning, some of the men of the 27th were still drunk or waking to hangovers. Ty said he dreamed about war that night. He woke at 2:30 a.m. December 8 to the sound of planes overhead, just as word of Pearl Harbor began to reach Manila. Some heard about the attack on commercial radio. Ty heard about it at breakfast. At first, he assumed it was a joke. When he found out otherwise from members of his squadron, they gathered their gear and rushed to Fort McKinley on the southern outskirts of Manila. He was back at his hotel by 8 p.m. preparing for duty at 4 a.m. From then on, he worked shifts at Nielson Field, Manila's commercial airport until war broke out, and just over two miles from Nichols field.

Although Ty had assumed he would fly a B-17—a heavy bomber known as the Flying Fortress—the 27th Bomb Group was to fly the SBD A-24, a lighter, more maneuverable bomber, also known as the Scout Bomber Douglas, the Banshee, or Douglas Dauntless. When Bjoring learned what they would fly, he wrote that it caused "the dismay of all pilots. ... But at the moment there were no airplanes of any kind to fly." Fifty-two of the planes were on a ship in a convoy bound for Manila, but the convoy was recalled to Honolulu after December 7.[65]

That Ty and the rest of the 27th, twelve hundred officers and enlisted men, had arrived in Manila before their planes violated a "cardinal rule" never to send a group to combat without its equipment. But when Ty and other pilots of the 27th Bomb Group left San Francisco in October, the country was at peace. These pilots didn't know it yet, but they were now in the infantry, members of what was called the "provisional army." The planes were rerouted to Brisbane, Australia, because it was too dangerous to send ships into Manila's harbor. Many of the planes available in the Philippines, especially for Filipino pilots, were hand-me-downs, "pathetically antiquated" planes U.S. forces no longer used. In her scrapbook the next night, Shy wrote, "Our anxious days began."[66]

Major Tom Dooley, Wainwright's top aide, saw a radiogram signed by MacArthur that morning saying a state of war existed. His reactions were subdued at first. Japan had struck multiple Pacific targets almost simultaneously. The fall of Wake Island and Guam cut communications between the Philippines and Hawaii. But it was the attack on Pearl Harbor and other islands that destroyed or severely damaged ships in the fleet that ended the possibility of sending aid to the Philippines anytime soon. For the next twenty-four hours at

U.S. bases around Manila, soldiers and officers were confused about what they should do, in part because of a delay in orders from MacArthur. Some of those officers were involved in the debate about whether to bomb Japanese bases on Formosa, as Taiwan was then known, before Japan could bomb targets in the Philippines.[67]

At 5 a.m., Brereton sought permission from MacArthur through his chief of staff to attack those air bases. He was told at 8 a.m. there would be no attack on Japan unless the Philippines were attacked first. When Brereton reported the decision to his staff, "there is no question that this statement came as a shock to most of the men in the room. ... Moreover, not to strike now meant throwing away the best chance they had at hitting the enemy on his own ground," Walter D. Edmonds wrote in an official Air Forces history of the war. Brereton got permission at 11 a.m. for a late-afternoon attack on Formosa. But it was too late. "The whole discussion is, after all, one of purely academic interest. ... There is but one sound explanation; and in the Philippines the personnel of our armed forces almost without exception failed to assess accurately the weight, speed, and accuracy of the Japanese Air Force."[68]

Particular blame falls on MacArthur, who was closeted with his staff for most of the morning, shutting out commanders who needed him to make decisions. Historians call it one of the most puzzling moments of his long military career. Author and military historian Bill Sloan argued that "without doubt, General Douglas MacArthur, the one person with full authority to control the situation, failed miserably to do so throughout the crucial period between December 8 and Christmas Eve." Authors Richard H. Rovere and Arthur Schlesinger Jr. made the same observation. From the first attack on, they wrote, "His mind seemed to be fixed on his grandiose plan for the defense of the entire Philippines. He refused to acknowledge the necessity of preparing for an eventual retreat" to Bataan and

Corregidor, the tiny, fortified island three miles off the tip of the Bataan Peninsula in Manila Bay. William Manchester, perhaps the most noted MacArthur biographer, wrote: "The key to the riddle is the General himself, and we shall never solve it, because, although those around him would recall afterward he looked gray, ill, and exhausted, we know little about his actions and nothing of his thoughts that terrible morning."[69]

The 4 a.m. message from Washington authorized MacArthur to put a war plan into effect that included using "planes to attack Japanese targets 'within range,' which clearly included enemy bases on Formosa." After sending planes up to look for Japanese attackers in the morning, commanders ordered them back to refuel and load bombs for the afternoon attack. The pilots at Clark Field, forty miles northwest of Manila, went to lunch. By 11:30 a.m., planes were parked wingtip to wingtip, despite orders from General Henry A. Arnold, the top Air Forces general, to disperse them if they weren't in the air.[70]

About an hour later, soldiers saw formations of high-flying planes. "They were pretty," one told writer John Hersey. "The boys thought they were Navy planes, except for one man who looked through the range-finder and saw but did not even have time to yell. First there was just noise. Then the men could see columns of dirt springing up, like a row of poplars in a picture, but moving, coming straight at them."[71]

Not everyone realized immediately that there had been an attack. Major Dooley, driving to Manila from Fort Stotsenburg, a few miles from Clark, thought he was seeing "tribesmen" setting brush on fire before he saw the formations of Japanese planes. "War came to the city in a slow, torturous haze." Residents of Manila went about their regular business for a while, then panic set in, followed by air raid sirens. "Life for the complacent people of Manila, an American city in denial, was about to change," author Peter Eisner wrote.[72]

After the bombings of Pearl Harbor and bases in the Philippines, the stories diverge. More than 2,400 U.S. service members and civilians were killed at Pearl Harbor. Troops stationed there were sent all over the Pacific to fight; many died. But after the attack, the Japanese fleet sailed away. Japan landed a few troops in the Philippines on December 10 and bombed bases again, destroying nearly all the planes in the country. An attack on Cavite Naval Base in the southern end of Manila Bay took out two U.S. ships and destroyed port facilities. The Navy had already moved most of its ships to Borneo and Australia, leaving six PT boats to defend Manila Bay. The main ground assault began December 22, when forty-three thousand Japanese troops landed at Lingayen Gulf, 200 miles northwest of Manila. More landed at Aparri, 360 miles north of Manila. "The missing dive bombers of the 27th Bombardment Group, for instance, could have had a field day with the Japanese transports that morning. But as it was the Japanese were only checked," Edmonds wrote.[73]

Like other pilots in the 27th, Ty worked shifts for the next few weeks. He and the others moved from the hotel where they had been staying to the officers club, where they could eat. Ty, Bjoring, Leif Kloster, and Alvan Ose, a flight school classmate from Iowa, were taken to the Air Forces headquarters at Nielson Field to work in air warning on the afternoon of December 8. Others blacked out windows, placed sandbags for revetments, and dug trenches. Bjoring wrote about being outside as a plane spotter, and Ty wrote about doing the same thing.[74]

Back at the officers club, they heard bombers overhead on December 9 about 2:30 a.m. "We all got out of bed and started to put on our clothes, and then the bombs started going off and the machine guns clattered and tracers flew in the air. It was all over before I could get my things and get downstairs. I was really shaking like a leaf and the nearest bomb hit was about a mile away," Ty wrote. He and Bjoring went to Nichols Field, which got most of the damage, and to Nielson, a little more than two miles away. They could see fires started by bombs. "Kloster and Ose were so scared they hadn't come out of the trench yet," which was probably an exaggeration.

Ty sent his parents a cable December 15 saying he was okay. It was the first they had heard from him since the December 8 attack. He went to a couple of movie matinees, "which were interrupted by an air raid each time." Ty and his friends bought extra food, including cans of pork and beans, tomato juice, and crackers. They adjusted to getting around at night during blackouts. Ty finally got his car off the ship that had arrived days after he did, and slept in it for two nights to protect it. He sold it on his twenty-second birthday, December 23, for $1,350. He wired $500 home and put the rest in a Manila bank. He cabled that news from Fort McKinley and wished everyone a Merry Christmas. His father wrote, in a letter Ty never received, that it was a good thing he had sold the car because his insurance didn't cover it during a war. The grounded pilots were still able to enjoy some niceties in Manila, however. "Life was not bad at all before we left there. ... We had ice in the ice box with Coca-Cola." Ty ate dinner with friends at the Army-Navy Club on his birthday, went to a show, took his watch to a repair shop, and listened to music at the Manila Hotel, where the band played "Happy Birthday" for him. He topped the day with an ice cream sundae. Six months on, he recalled the day: "I will remember it as the last bit of social life before Bataan."

He sent a second cable to his parents in reply to one they sent him. They wished they could have sent cables more frequently, but at $5.92 each, they couldn't afford it. "Of all the things that have ever been given to me, these two cables have meant more than anything I have ever had as a gift," Shy wrote.

Hans and Shy drove to Alliance, Nebraska, after they heard about the attack on the Philippines. Her brother and his wife,

Bob and Grace Cobb, didn't know where their son Chan was. "All we did was listen to the radio and talk about how treacherous the Japs were," Hans wrote. Shy wondered in a December 9 letter if Ty had been able to bomb the attackers. Chan, who was in the Navy, cabled on December 12 that he had been at Pearl Harbor during the attack but was fine.

"Meanwhile, the stories of American victories that appeared almost daily in stateside newspapers during this period were often gross exaggerations or complete fabrications," Sloan wrote. Newspapers in Nebraska, including those Hans and Shy would have read, seesawed between reports that were optimistic and those that must have terrified them.[75]

On December 15, the *Lincoln Star* ran a column headlined: "Japs Monday Found the Going Increasingly Tough." The writer added, "American aviators were getting their initial baptism in the sky, and were giving a truly glorious account of themselves." A few pilots went up in aging planes that survived the December 8 attacks and bombed Japanese ships, but Ty was not among them. Another column the same day said Americans were "holding our own ... pending the arrival of reinforcements."[76]

On December 16, headlines revealed the ghastly toll at Pearl Harbor for the first time. Three days later, the *Star* reported about "a marked increase" in land and sea attacks on Luzon. Amid reports that American submarines were successfully attacking Japanese ships was the news that Hong Kong was about to fall to Japan. On the 20th came the news that Japan had landed troops on Mindanao, the large southern Philippines island.

By December 22, the *Star* banner headline told readers "80,000 Japs Attack Philippines," followed by a positive spin in the subhead: "U.S. and Filipino Troops 'More than Hold Their Own' Against Enemy." That bravado disappeared two days later, amid scarce details. The *Star* reported that MacArthur and his staff were in the field under a stark headline: "JAPS ADVANCE / Crisis Near in Headlong Philippines Battle / Imperiled Manila May Be Declared an Open City." On Christmas Day, the paper reported that military officials in Washington said Americans should expect more bad news and the possibility of losing the Philippines. Japan's "invasion horde" was holding positions fifty miles from Manila.

MacArthur's initial war plan was to defend the entire country, 7,100 islands stretching 1,150 miles from north to south, with his ground troops and a fleet of fifty PT boats. He ended up with nine. Shortly after the December 8 attack, it became clear that plan was unrealistic, yet MacArthur hung onto it. He wasn't the only one. As Major Dooley watched Japanese bombers overhead December 14, he wrote: "They, for the present, have air superiority and know it. They seem to bomb at will, but I know what our air corps has at the time. It will be quite different soon. The Japs will rue the day they started this fracas." A week later, Dooley was back in Manila, where he attended a party at the Manila Hotel. "Much fun, place quite crowded. No evidence of war except officers in uniforms with pistols."[77]

Others saw incompetence that cost lives. After his troops in northern Luzon were overwhelmed by Japanese attacks, MacArthur put War Plan Orange-3 into effect the morning of December 23. Critics said he should have done so days earlier. "The capital and the General lived in a world of fantasy for two crucial weeks," MacArthur biographer William Manchester wrote. The plan called for troops to fall back into the mountainous jungle of the Bataan Peninsula northwest of Manila and fight until reinforcements arrived. "This was sheer nonsense and MacArthur knew it," author Mark Perry wrote. Military officials who had reviewed the plan years earlier said to assume the quick rescue of troops by the Navy "would be literally an act of madness," Perry added. A more realistic plan for a naval rescue would be two or three years.[78]

MacArthur declared Manila an open city, hoping to spare damage and civilian lives and to avoid having to split his forces to defend it. That move temporarily confused Japanese General Masaharu Homma, the leader of the invasion. He expected little resistance in central Luzon and that the decisive battle would be fought in Manila. Homma had been advised that Bataan, "being a simple, outlying position, would fall quickly." The few remaining U.S. planes on Luzon were flown to Australia, six thousand miles to the southwest. American and Filipino troops in the Philippines, as well as Wake Island and Guam, didn't know it yet, but they were on their own, "declared expendable in the event of a heavy Japanese offensive," Sloan wrote. Instead, the Allies would focus first on defeating Germany.[79]

"Bataan, before it became hell, was kind of an unfinished paradise," Hersey wrote, calling it "incredibly wild." It is thirty miles from north to south and twenty miles wide, with a mountain range down the middle. The peninsula has beaches facing Manila Bay and cliffs facing the South China Sea. There were cold-water streams and ponds, huge banyan and mahogany trees, and plenty of wildlife.[80]

MacArthur and his family joined the Christmas Eve mass evacuation to Bataan and Corregidor, which caught Ty and many other soldiers by surprise. "I worked in the morning and during the afternoon," Ty wrote. "We were thinking about going into town for some things. We called for a car, and they said, 'All transportation was being held ready for an evacuation.' It was the first any of us ... knew about it, so we thought we'd better be finding out. ... I started packing, shaved, took a bath and got ready to leave."

Ty had a suitcase and laundry bags ready at 5 p.m. when he and his friends "piled on the truck and to the port area we went.

As we left town, the kids were giving us the 'V' for victory sign. We stayed on the docks until about midnight loading the boat. There was a full moon and everyone was scared of a raid as it was quite light. It was really a mess. Ammunition, food, files, equipment—what a mess!" After his boat nearly collided with another boat, he spent a restless night at Corregidor Island just south of the peninsula, fearing a possible air raid, and docked on Bataan Christmas morning.

For others it was the "wildest most fantastic day," as Edmonds wrote. Colonel David L. Hardee described moving to Corregidor as Japan was shelling the port: "I shall never forget that Christmas Eve. ... The scene would have been picturesque had it not been so ghastly and our plight so desperate. The sky was lighted for miles by burning supplies of gasoline and oil from Fort McKinley, Nichols Field, Pandacan, and other strategic points. The flames licked the sky and lighted Manila Bay. ... The whole thing looked like a billion dollar Christmas tree." U.S. forces destroyed oil and other supplies to keep them out of Japan's hands.[81]

Many soldiers in Manila went to Bataan by truck, driving north, then taking a sharp turn south where the bay was narrow enough for bridges. Many of those trucks were empty, an episode soldiers would "remember with legitimate bitterness" as they starved, author Winston Groom wrote. "The peninsula itself was bedlam," Toland wrote, with trucks jammed on roads with fleeing civilians, and Japanese troops not far behind them.

Under optimal conditions, it would have taken two weeks to move food and supplies to Bataan. The late decision to put WPO-3 into effect meant shortages. MacArthur had sent trucks with supplies farther north on Luzon where he expected the main fight would take place. They all disappeared, contributing to a shortage of both food and vehicles. A warehouse with 50 million bushels of rice in central Luzon was abandoned. It could have fed the defending armies for more than four years.[82]

Meanwhile, Bataan had enough food for forty thousand soldiers for a month. But there were eighty thousand soldiers, thirteen thousand of them Americans, and twenty-six thousand Filipino civilians. On January 5, 1942, MacArthur ordered half rations for everyone. He made his only visit to Bataan five days later, temporarily boosting morale, but drawing withering criticism from enlisted men and officers alike for remaining on Corregidor, which most of the troops regarded as safer and better supplied with food. They called him Dugout Doug. Fighting in Luzon intensified as troops finished withdrawing into Bataan. "What a shame," one soldier recalled after the war, "these high-paid Air Force pilots down there fighting as infantry."[83]

On Christmas Day, Ty wasn't sure what to do. He ate a can of pork and beans and drank a can of tomato juice. An officer needed men who could drive trucks to help move supplies being removed from Fort Wint, on a tiny island in Subic Bay. The Army planned to blow the fort up the next morning. Ty and three others volunteered, and more trucks went later. When he got to Olongapo, on the mainland across from Wint, "everything was a mess, and they didn't know what they were doing." He took charge. "Some of the crazy officers there wanted to leave ammunition behind so they could take personal belongings such as camphor chests filled with linens, etc. I told them ammunition has priority and we would get personal stuff last. They were even going to leave a couple of search lights behind. Some people are sure smart. There are so many dumb head tricks that have been pulled in this war that it makes me mad every time I think of them."

He found a .45 revolver and a Browning automatic rifle while he was at Olongapo. Someone stole the revolver the next day, and the rifle was too heavy, so he gave it away. By dawn, everything of importance was off the island, and Ty spent most of December 26 putting guns in place at Olongapo. Late in the day, he returned to Mariveles, the southernmost city on Bataan and the main port, to retrieve his belongings, which weren't there. Exhausted, and with only the clothes he was wearing, he slept in the back of a truck. He and his pals, Leif Kloster and Alvan Ose, who had been part of the mission, were ordered to rejoin their unit, which they eventually found, along with their things. They set up camp at what Ty called "a swell place." There were lots of trees and a creek for swimming. They built a dam to make the swimming hole deeper. They were soon to be moved to the southern island of Mindanao, and from there to Australia where their planes were. "As you know, the move never came for reasons I won't write about," he later wrote in his diary, one of his early comments on what he, and others, considered poor planning. "Hope responsible party gets got after the war." He set up a tent over a foxhole, got a cot, and hung a sign out front that read "Hotel Hyannis."

In letters, families tried to project normalcy to boost their soldiers' spirits. Read many years later, some of what they wrote seems distinctly odd. Ty, of course, never saw the letters returned to his parents. Hans, a World War I veteran, then 53 years old, wrote in December just after the war started: "Wish I could be right there with you & could be one of the crew on your bombing trips. We would give those Japs hell. As I can't, you will have to do my part for me," promising he and Shy would do what they could to make sure soldiers had what they needed. In a January 19 letter, Shy thanked Ty for the Christmas gifts he shipped before the war started, then listed other things she would like when he found time to shop. She wrote that radio news reported all the work soldiers were doing to repulse Japanese landings on the islands. "You are surely in it," she wrote, adding that she always wore the wings pin he had given her when he graduated from pilot training. In an early

January letter, Hans wrote, "We know you are up there flying and you will let us know how you are as soon as it is possible." He said they weren't worried and knew Ty and the others were well fed and clothed. One of Ty's teachers from Kearney wrote to Hans and Shy to praise his heroism, but like his parents and other Americans, she was too optimistic: "What a wonderful experience for a young man!"

Around this time, Ty was appointed mess officer by Lieutenant W.G. Stirling, commander of the 17th Squadron. As part of the gig, Ty drove a truck around Bataan looking for food. On New Year's Eve, Stirling told Ty to find fresh meat for a New Year's Day dinner. "I went down the road and hit a bombing at Lubao and saw my first casualties of the war – most of them natives. They [Japan] had hit an ammunition train. ... there were really fireworks when I went past."[84]

Hearing there was plenty of meat in Manila, he drove there, arriving after dark at the Leonard Wood Army-Navy Club. He spent a restless night with "the sound of all the gas and oil in the vicinity of Manila going up in smoke," and was up before dawn. Others made runs to Manila for food, but they started early and left the city before dark on December 31. A few celebrated that night. Major Damon "Rocky" Gause was sent to Manila from Bataan to retrieve radio equipment and was invited to a party at the Manila Hotel. "Everybody danced that night, young and old." They drank too much, he reported. "This was more than an ordinary New Year's party. We were commemorating the passing of an era as well as a year."

The port was a wreck, ruling out the most direct route across Manila Bay. Both Gause and Ty began the new year of 1942 driving back to Bataan. Gause left before dawn. His convoy of four cars and trucks was the last to get to Bataan, as bridges were blown up to delay the Japanese army. Ty now had fifteen Filipino passengers and his truck—loaded with meat, eggs, ham, livers, and more. He left later than Gause and soon discovered a blown bridge. Ty sent the truck back to Manila, and he and the soldiers took a canoe across the river, where they found an abandoned truck, which Ty hot-wired. After a short distance, a local boy jumped on the truck with a warning: "The Japs are in the next town up the road about seven kilometers blowing the hell out of everything with tanks." For safety, they broke into smaller groups. Ty and three flying cadets walked into a small town, only to be met by a group of armed men who were pleased to see American, rather than Japanese soldiers. After feeding them, the villagers transported them across the bay in a small boat. Ty saw sharks in the water and dive-bombers overhead. He hitchhiked back to his squadron, "with one less truck than I started with." It was a day, he said, he would remember for the rest of his life.[85]

The flying cadets' wardrobes included fur-lined leather
helmets, goggles, leather jackets, parachutes, and silk
scarves. Family photo

When Ty was in flight training in Oxnard California, in the
fall of 1940, he often joined a friend, who was in graduate
school in Los Angeles, for frenetic weekends of movies,
dinner dances, and trips to the beach. Family photo

Ty took a supply of Hotel Hyannis stationery with him to pilot training. It lists him, along with his parents, as proprietors. In this January 10, 1941, letter, he began several pages about the thrill of flying at Randolph Field, near San Antonio, Texas

Chapter Six

Running out of Food and Time on Bataan

"You are well aware that you are doomed. The end is near. The question is how long you will be able to resist."

—Japanese General Masaharu Homma, commander of the Philippines invasion, in a message to MacArthur, January 10, 1942

Ty and the rest of his unit settled into a camp near Limay about twenty miles up Bataan's east coast from Mariveles. They named the camp Sylvan Woods, primarily because of the trees and stream. Otherwise, it was nothing like the mythical realm of Greek and Roman gods. There were some comforts, however. A chaplain held services on Sundays. A radioman in the tent next to Ty's was able to tune in KGEI, a San Francisco radio station that broadcast on short wave. It was the only U.S. station that could be heard in the Pacific. And they waited for word their planes had arrived in Mindanao, where they would be taken to claim them.[86]

The soldiers were happy to hear news, but advertisements for food were tough. More often, what they heard in camp were rumors, including that Russian planes had bombed Japan.

If information was hard to come by in Bataan, news stories back home were sometimes more optimistic than they should have been. Shy pasted one headline, dated February 1, in her scrapbook: "MacArthur Is Holding Off the Japanese." Next to it she wrote, "No news from you Ty, but we know you are doing your duty some place."

Now in the "provisional army," members of the 27th Bomb Group were run through a few drills to make up for their lack of infantry training. Ty said his unit was "schooled in infantry by a lieutenant colonel." Lieutenant Colonel David L. Hardee was the executive officer of the Provisional Air Corps Regiment, made up of two thousand men. "We found them ready, able, and willing. It was a pity to take all the good pilots, bombardiers, aerial gunners, and mechanics of many years' training ... and not be able to use them to the fullest of their special skills," he wrote in his memoir. Not everyone was as impressed with the training. An officer in Ty's unit praised the pilots and others but said they were only "remotely acquainted" with their rifles.[87]

As mess officer for the 17th Squadron, Ty went to the quartermaster for rations every third day. At first there was plenty of canned food, but not much variety. But the shortages were stark, according to a January 3 inventory; the Army had enough food to feed one hundred thousand men on Bataan for a month. MacArthur ordered rations cut in half two days later. "The half-ration, containing about 2,000 calories ... was obviously inadequate to the needs of fighting troops who had to work as much as twenty hours a day, under the most difficult conditions and in the worst kind of climate and terrain. Fortunately many of the men had accumulated food during the withdrawal and this supply was used to supplement the meager diet," the report said.[88]

When they could no longer feed their mules and horses, the Army slaughtered and ate them, although some could not, despite their hunger. Soldiers hunted dogs, which "tasted like lamb," iguanas, and monkeys. One colonel wrote about his preferences: "I can recommend mule. It is tasty, succulent and tender—all being phrases of comparison, of course. There is little to choose between calesa pony and carabao. The pony is tougher but better flavor than carabao. Iguana is fair. Monkey I do not recommend. I never had snake." (Carabao is the Filipino water buffalo.)[89]

"The life expectancy of anything that walked, crawled, or flew on lower Bataan was practically nil," as Dyess put it. Groom wrote, "Soon everyone's clothes hung off them like a scarecrow, that is, what was not already in tatters or rotten from the harsh jungle fighting." The average American, who weighed 170 to 200 pounds in January, was down to 150 within a month. Ty weighed about 170 pounds in January. By February, the soldiers' rations were down to a fourth, and by March to three-eighths. By April 1, the surgeon to Luzon commander General Edward P. King wrote that the combat efficiency of the troops in Luzon "was rapidly approaching the zero point."

Poor diet wasn't the only problem. "The absence of mosquito netting, shelter halves, blankets, and sun helmets was as serious as the shortage of clothing. The physical deterioration of the troops and the high incidence of malaria, hookworm, and other diseases were caused as much perhaps by the lack of proper protection against the weather and the jungle as the unbalanced and deficient diet," Army historian Louis Morton wrote. By the end of February there was a critical shortage of several drugs, most important the antimalarial drug quinine. Two of Ty's buddies came down with malaria at about that time, as he would some months later.[90]

By January 7, Ty's unit had moved to the secondary front, close to Orion, about forty miles north of Mariveles and near some rice fields. He pulled guard duty, dug a spider hole (similar to a foxhole, but covered and set up so a soldier could see out to fire his gun), laid more barbed wire, helped to build a machine-gun nest, and was taught how to throw a hand grenade. The sounds of machine guns, mortars, and rifles made sleep difficult. Japanese forces were getting closer, and U.S. and Filipino troops stationed near Guagua, a few miles northeast of the peninsula, began to withdraw. As they moved south during fierce fighting, they blew up bridges to make it difficult for Japanese soldiers to follow them. Approximately half of the Filipino soldiers left during the retreat and went home. Tom Dooley, Wainwright's aide, wrote in his diary they were "always falling back—but what can we expect—untrained troops—few automatic weapons—and no air support."[91]

Ty was officer of the guard once a week, then every other week. He was a courier and took on jobs others didn't want. The pilots waited for help from home, hoped to be taken south so they could fly, and worried they would be captured. "This whole business was and is a sore spot to me, so if I am cynical, you know why," Ty wrote in a rare criticism. He wasn't alone in his resentment. By the end of January, Dooley had begun to lose faith that aid would come in time: "I believe the War Department looks on us as a brave band and that is all. We do not really count in the big picture. ... We must be sacrificed for the fulfillment of the carefully planned movements of the 'giant efforts.' If I do live thru this ... I am going to get the true picture to the men in Washington."[92]

Ty was glad to be part of a group sent to Air Forces headquarters near Manila Bay, which was farther from the front line, for about a week as observation and fire-watch officers. The job was boring and the food, mostly cold, was "lousy." They ate two meals a day, including one breakfast of salmon with onions and pickles. They were in camp during the day and on the beach at night. One night in late January, Ty wrote, "It was after dark when I got to my O.P. [observation post], and bang! Off went a 155 (one of our artillery guns) ... and then the fireworks started—machine guns, rifles, tracer bullets were going every which way. A bullet whizzed by my head, and

I hopped into an armored car." He and the others assumed the United States was about to surrender. Later, they learned it was their own troops conducting a test, but no one knew, and Filipino soldiers had returned fire. The fracas lasted about fifteen minutes.

The day after Ty's escape from Manila, Japan took control of the city. Japanese soldiers lowered the American flag, which one soldier stomped on, and raised the Japanese flag. General Homma moved into MacArthur's suite at the Manila Hotel and hung an enormous Japanese flag on the balcony that could be seen from Corregidor with binoculars on a clear day. As a tall, Western-educated man, Homma was seen as almost foreign by his colleagues. He told his superiors that capturing Manila was the main event and that soldiers on Bataan were the remnants of a disorganized flight who could be quickly defeated. He was forced to send some of his troops to Java, then discovered MacArthur had nearly an entire army on Bataan, more soldiers than Homma commanded. Japanese forces installed a puppet government, imprisoned civilians from the United States and other Western countries at Santo Tomas University, settled into the internees' former homes, took over the news media, seized cars and trucks, looted stores and warehouses, and freed twenty-five thousand Japanese prisoners.[93]

Japanese planes dropped leaflets signed by Homma over Allied troops January 10, not realizing MacArthur had chosen that day to make his only visit to Bataan. The leaflets warned of imminent defeat: "You are well aware that you are doomed. The end is near. The question is how long you will be able to resist. You have already cut rations by half. I appreciate the fighting spirit of yourself and your troops who have been fighting with courage. ... You are advised to surrender." MacArthur ignored the message, as did most soldiers. Author Peter Eisner quoted the general's response: "Every foxhole on Bataan rocked with ridicule that night." Some used the leaflets as toilet paper.[94]

Without enough food, soldiers on Bataan lived mostly on the belief that reinforcements would arrive before it was too late. Roosevelt barely mentioned the Philippines in his speech December 8, 1941, asking Congress to declare war. But a statement he issued three weeks later brought wasted hope. He praised Filipinos' "gallant struggle against the Japanese aggressor" and promised that the Allies would provide resources so their "freedom will be redeemed." His speech brought joy to soldiers and civilians in the Philippines. Roosevelt and others in Washington quickly realized he had seemed to promise immediate help. It fell to Roosevelt's press secretary, Stephen Early, to tell reporters that they had read "too much of the immediate rather than the ultimate into the President's statement."[95]

A few days after his visit to Bataan, in a message to his troops, MacArthur, too, promised aid was on the way. He was careful not to state the date of arrival, but "the hungry men, grasping eagerly at every straw, assumed that it would come soon," Manchester wrote.

On January 30, the ninth anniversary of Nazi rule, Hitler spoke of ridding Europe of all Jews. Two days later, the first U.S. air attack of the war inflicted severe damage on Japan's air and naval bases on the Gilbert and Marshall Islands, south of Wake and Guam and near New Guinea. At the end of February, the United States was defeated in the four-day Battle of the Java Sea.[96]

At Air Forces headquarters, Ty lived in a large pyramid tent with his buddies, Leif Kloster, George Davis, and Bob Bjoring. They "made a fine foursome, after we taught Bjoring how to play bridge." In their free time, they made boards for games. Ty played poker and chess, but he preferred bridge and cribbage, which has long been the unofficial Kokjer family game. He thanked his parents for the playing cards: "You couldn't have given a better present," adding they were practically worn out when he had to abandon them after the surrender. They talked about food, what had gone wrong in the war, their colleges, their hometowns, and their lives before the war. He and others built a bomb shelter, tables, and shelves and raised their beds off the ground to prepare for the rainy season.

For now, Ty said, they were eating as well as anyone on Bataan. At breakfast, he might eat six or seven hotcakes. Later, the hotcakes were watered down, and eventually, they ran out of flour, meaning there was no bread. For dinner, and sometimes for breakfast, they had rice and vegetables. Sometimes they had salmon, sardines, or carabao. After the first few weeks on Bataan, "the search for food assumed more importance than the presence of the enemy to the front. Every man became a hunter, and rifle shots could be heard at all hours far from the Japanese lines," Morton wrote.[97]

Hans and Shy must have known that mail service to the Philippines had been cut. But they still looked for letters from Ty. In a February 14, 1942, letter, the American Red Cross informed them that Ty had not been reported among the "lost, missing or injured," as of February 7. His parents fervently wished he had made it to Australia, although had he reached Australia, they surely would have heard from him. Radio news reported there were too few planes for the pilots, and his parents then assumed he was fighting with ground troops. A newspaper in Kearney wrote that Ty was able to let his parents know he was safe—at least for a while. If Ty did nothing, the Army deposited a set amount of his monthly pay in the Hyannis bank. If the amount was different, even by a few pennies, they knew he had made the change and was safe. Soldiers were sometimes able to get messages to radio operators, and in late March, Ty got a radio cable out. Shy called it "a red-letter day" and happily shared the news with family and friends. It would be his last message to them for more than a year. In the months that followed, Hans and Shy and other relatives wrote letters seeking information about Ty.

In a February 23 fireside chat, Roosevelt finally made it clear the Philippines would not get help anytime soon. Most Nebraskans had little choice but to stay home and listen to the president on their radios—a statewide blizzard had closed many roads. Listeners had maps of the world in front of them, as Roosevelt had suggested. Newspapers in the Nebraska towns of Hastings, Grand Island, Columbus, and many others across the country printed maps that day with the same headline: "Clip This Map ... Study It as You Hear Roosevelt Tonight." The text under the map continued with directions for its use: "Have it with you when you listen to Mr. Roosevelt at 9 p.m. Central War Time. Learn what this war means to our America." (What was once called war time is now known as daylight saving time.)[98]

Roosevelt pointed out that Japan was close to its southwest Pacific targets, while the United States was thousands of ocean miles away. Japan, he said, had taken control of Hong Kong, Singapore, Wake, and Guam almost simultaneously. The retreat to Bataan and Corregidor had long been the strategy, Roosevelt said, adding, "It is that complete encirclement, with control of the air by Japanese land-based aircraft, which has prevented us from sending substantial reinforcements of men and material to the gallant defenders of the Philippines." Doing so "would have been a hopeless operation."[99]

As Roosevelt spoke, a Japanese submarine shelled but inflicted only minor damage at an oil refinery near Santa Barbara, California. That attack and the blizzard overshadowed the president's speech in front-page headlines of the *Nebraska State Journal* the next day, although the paper printed the entire text on page three. A front-page news story quoted the president as saying the country had "been compelled to yield ground" and that the Allies "are committed to the destruction of the militarism of Japan and Germany." He promised the United States would take the offensive soon. Roosevelt praised MacArthur and his men for holding out on Bataan longer than expected. They had "exceeded previous estimates and he and his men are attaining eternal glory therefore" and making Japan pay a high price for its advances.[100]

The day before his fireside chat, Roosevelt privately ordered MacArthur to leave the Philippines for Australia, but gave him some leeway on the departure date. Roosevelt made it clear the general was leaving under his orders, to tamp down the scorn many would heap on MacArthur for abandoning his troops. Allowing the famous general to be captured would have been a political disaster. The general ordered Wainwright to come to Corregidor on March 10 and handed him command of the Philippines. MacArthur left the next day, arriving in Australia after a harrowing, weeklong journey by sea and air. Even though historians have found fault in his actions from December 8 to his departure from the Philippines, some feared his absence would cause soldiers to stop fighting.[101]

Hersey wrote a sympathetic appraisal in his book that was published in the spring of 1942. Americans "began to expect things of him that he could not do. ... When he went to Australia, the people heaved a sigh of relief and sat back to watch the Pacific tide turn, believing that this man, rather than their own efforts, would work the miracle." Some soldiers felt abandoned. "It was as if the star quarterback had been pulled from a championship game," Charles Underwood Jr. wrote about his soldier father's likely reaction.[102]

Other soldiers had few compliments for the general. Before leaving Bataan, MacArthur had issued 142 communiqués, "all vividly written and making wonderful reading," with 109 that mentioned no one's name but his. "It was always MacArthur's men, MacArthur's left flank, MacArthur. On Bataan they choked on the sound of the name," author Gavan Daws wrote.

MacArthur portrayed himself as singlehandedly standing up to the Japanese. "He did so through carefully crafted communiqués—many of them erroneous—that he personally labored over for up to two hours each day," author James M. Scott wrote. There is little doubt that Americans at home were fans. Cities named streets after him, many boys were christened Douglas, and there was even a dance, the MacArthur Glide.[103]

News of MacArthur's departure from Bataan was not revealed until March 21. The *Pittsburgh Press*'s breathless report was typical: "Here's Thrilling Story, of Dash by MacArthur," adding that the details of his "intrepid" escape had been announced in Melbourne, Australia, where MacArthur set up his new headquarters.

In the days before the general's escape, papers across the country were serializing a biography by journalist and author Bob Considine called *MacArthur the Magnificent*. They ran optimistic stories about victories in Bataan, such as "MacArthur's Bataan Stand Ranks with Alamo, Thermopylae." Papers reported there had been no changes in Bataan, that MacArthur's soldiers had pushed the defensive line forward five miles, and that "MacArthur's great feats are possible because of the fortification of Corregidor and his almost impregnable mountain defenses in Bataan."[104]

In fact, Homma had called a temporary halt to his offensive in mid-February, beginning an "ominous calm" that lasted more than a month. Japan had expected to conquer the Philippines in fifty-five days. Homma, who thought he could win in a month, had "failed miserably." During the lull, Japan sent reinforcements and supplies. That Homma had not succeeded by the end of February ruined his already tenuous career. After the United States surrendered, Homma was recalled and "cashiered," a term used mostly by the military when someone is dismissed or demoted for an infraction of the rules.[105]

During that lull, Ty said he stayed in camp most of the time. "Bombers had been flying over and we had been running for our fox hole at least six or eight times a day and sometimes at night you could hear the planes coming, that certain whistle the bombs have, and then the explosion. Although we were in no target area to speak of, we didn't trust the Japs' accuracy." He had seen two enemy planes shot down a few days earlier: "One going down smoking and the other spinning in flames. It didn't make me feel happy. I just thought that the crew of those planes have families somewhere that will never be all together again because of what I had just witnessed." He got out to visit friends, including one who had been at the Battle of the Points on the peninsula's southwest coast January 23 to February 1, a hard-fought, short-term win for U.S. and Filipino troops. Ty also visited friends stationed at Bataan Field east of Mariveles.[106]

On March 19, Homma dropped his next message suggesting surrender to Wainwright and U.S. soldiers on Bataan. The notes were stuffed inside hundreds of beer cans. Homma said Wainwright had done his duty, and now it was time to surrender. The Japanese general complained that Wainwright had not replied to a previous message, promised to follow international law regarding the treatment of prisoners of war, and said if there was no reply by noon March 22, Japan will "consider ourselves at liberty to take any action whatsoever." Wainwright's reply was contemptuous: "The bastards could at least have sent a few full cans of beer." Although Ty wasn't a drinker, most soldiers probably shared Wainwright's reaction. Other messages dropped during that time urged Filipinos to kill Americans. The messages included tickets for surrender and promised humane treatment of Filipinos when Japan won.[107]

By then, all U.S. forces that had been pushed ever south were nervous. "Bataan was a hopeless hell where everything was bad except the will to live, the memories of home (as torturous as they sometimes could be) and the ever-dimming hope that the great country we represented would somehow find a way to help us," Wainwright wrote, adding, "malaria, another ruthless enemy, hung over us like a black cloud, enveloping a land of men whose bones were lumping through their tightly stretched skin."[108]

Some in the military still characterized Japanese soldiers and equipment as inferior. One colonel said they were "distinctly fourth-raters," calling it a "charitable estimate," which infuriated Washington officials. Secretary of War Henry L. Stimson said at a press conference that the Japanese were better trained and equipped than some forces under American command. "Unfortunately they were good at flying, good at shooting, good at planning, good at landing, good at engineering, good, in fact, at nearly everything," Hersey wrote. Japan had conquered a fourth of the world's surface and was free to roam nearly the entire Pacific Ocean. American generals regarded Japanese soldiers as good at following orders, better on offense than defense, more ferocious than German soldiers, but not as good as American soldiers at improvising, something that soldiers would soon observe firsthand. (Mark Perry wrote Japan controlled a fourth of the world; William Manchester wrote Japan controlled a seventh.)[109]

With fresh troops and plenty of supplies, Homma resumed his attack on March 24. Rations for U.S. soldiers were at one-third or less, "poorly balanced and very deficient in vitamins." Without more supplies, "the troops will be starved into submission," Morton wrote. It was only at the end of March that officials in Washington learned how many people, military and civilian, were stranded on Bataan, nearly 110,000 and that they had only enough food to last until April 15.[110]

Ty's pilot training was delayed for four months in 1941 when he broke his leg in a car accident. After his cast was removed, Ty, left, spent time recuperating at the beach at Galveston, Texas, with is cousins Chan and Bobby Cobb and a friend. The photo is not labeled, so the identity of the second man isn't known. Family photo

It's not clear where this photograph of Ty in his Army
uniform was taken. Given what appear to be cranes in the
background and the shiny wings on his uniform, it was
likely taken in San Francisco before he got on the ship to
Manila. Family photo

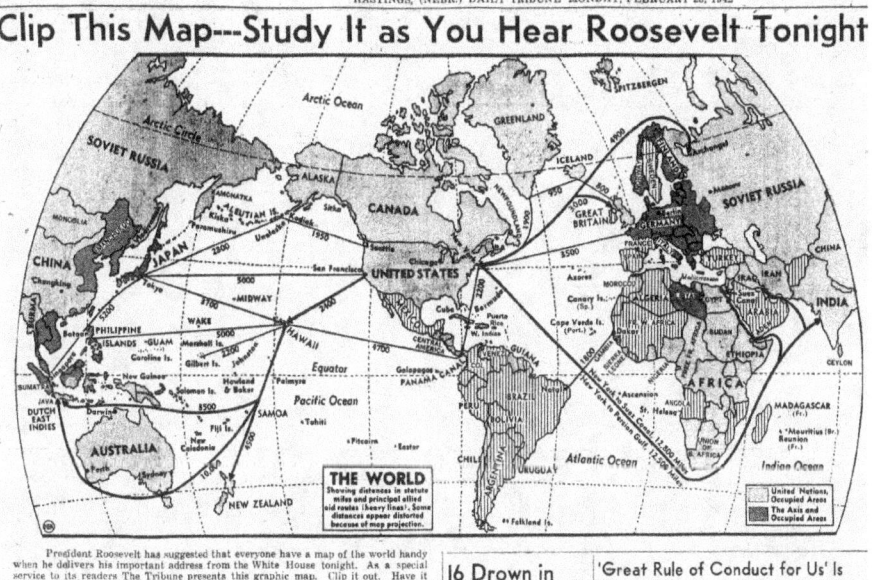

Clip This Map---Study It as You Hear Roosevelt Tonight

President Roosevelt has suggested that everyone have a map of the world handy when he delivers his important address from the White House tonight. As a special service to its readers The Tribune presents this graphic map. Clip it out. Have it with you when you listen to Mr. Roosevelt at 9 p. m. Central War Time. Learn what this war means to our America.

16 Drown in River Tragedy

'Great Rule of Conduct for Us' Is In Washington's Farewell Address

WASHINGTON, (UP)—The sen-|or a remote relation. . . Why

Newspapers across the country printed world maps February 23, 1942, at the urging of President Franklin D. Roosevelt. In his fireside chat that night, he said Japan's control of key Pacific Ocean islands meant reinforcements would not arrive in the Philippines any time soon. Map courtesy of The Hastings (Nebraska) Tribune

Chapter Seven

From Horrible to Horrific

"With a minimum of aid Bataan could have been held."

—Major Thomas Dooley, April 9, 1942

Ty and his friends tried to keep their spirits up, even though they knew their situation was dire. On April 1, he pulled an April Fool's joke, which he didn't describe. "And someone said I should be given the Distinguished Service Cross for even being able to have the morale to remember April Fool's Day." For a young man who expected to spend two years in the military, then return home to live a middle-class life modeled by his parents, the past four months were not what he had planned. Nor was it what his family had assumed. Ty's mother wrote to him about plans to visit Manila in 1942, halfway through his two-year tour of duty.

After marching ever south on Bataan, Homma selected April 3 to begin his final offensive. It was Good Friday and a special day for the Japanese, the anniversary of the death of Emperor Jimmu, by legend the first ruler to sit on the imperial throne. On Easter Sunday, April 5, Dooley visited Bataan with his boss, General Wainwright. They ate breakfast with Luzon commander General Edward P. King and his staff. "In my eyes 'DEFEAT' was written all over them," Dooley wrote in his diary.

Homma hoped to succeed by April 29, Emperor Hirohito's birthday. Instead, he had all but won by April 7. "The story of the last two days of the defense of Bataan is one of progressive disintegration and final collapse," Morton wrote about the chaos from April 7 to 9. "In two days an army evaporated into thin air."[111]

American and Filipino soldiers had held off the Japanese invasion of Bataan far longer than anyone had anticipated. But now, the Army was out of food, and King's soldiers were suffering from malnutrition, beriberi, dysentery, malaria, and other diseases. Only twenty-seven thousand of them were "combat effective," and three-fourths of them had malaria. Despite orders from Roosevelt and MacArthur to attack the Japanese invaders and never surrender, King knew his soldiers would have been annihilated had he obeyed orders.[112]

In his last call to Wainwright, who was on Corregidor, at 3 a.m. on April 9, King didn't mention that the surrender of Bataan was already underway. King expected to be court-martialed once the war was over. But his peers were sympathetic. "I know the real person that Gen. King has always been. He did it to forestall more deaths and suffering. He could see the cause was hopeless," Dooley wrote. "With a minimum of aid Bataan could have been held." Wainwright later wrote: "If there is anything worse than a battlefield that shakes with explosions and the cries of men it is one that becomes mute and dead and just sprawls there broken and exhausted. That was Bataan on the night of April 9, 1942."[113]

Ty began to realize the end was near, as he wrote later in his diary: "April 9th was the climax to a drive that the Japs had been making for a number of days before. They had broken through our main line of resistance the day before and really upset things. We all felt quite secure until evening when reports came in that the line had broken again. Then when Corregidor started sending shells at them from in back of us, we knew how close they were as we knew Corregidor's range."

Later that day, King met with a Japanese colonel, laid down his pistol, and surrendered the largest number of troops under U.S. control ever, seventy-eight thousand U.S. and Filipino soldiers. About twelve thousand of the soldiers were Americans. King asked that his men be treated well. The Japanese officer across the table replied that they were not barbarians. That would be proved wrong over and over again in the next three and a half years. As author Bill Sloan wrote, "The orgy of cold-blooded barbarism that followed would forever disprove that assurance."[114]

After the surrender, Roosevelt withdrew his order to attack: "Am keenly aware of the tremendous difficulties under which you are waging your great battle. The physical exhaustion of your troops obviously precludes the possibility of a major counter attack."

Edgy, malnourished, and sick soldiers waited on that hot day to see what would happen next. Ty hadn't slept much the night before. For the soldiers, whose lives were about to go from horrible to horrific, an earthquake was a cruel joke. "What a night!" Ty wrote later in his diary. "Ammunition being blown up, the sky red from fire and black from smoke. And to top it off, we had an earthquake to go along with it." Others also reported scenes of ammunition dumps being destroyed: "The whole southern end of Bataan was like a volcano erupting, with white-hot metal fountaining from exploding bombs and shells, colored flares rocketing everywhere, snaking through the air like crazed disintegrating rainbows," Gavan Daws wrote.[115]

When the surrender came, Ty had a pack ready. Stuffed inside were his orders, some clothes, a blanket, a half tent, medicine, a water purifier, emergency rations, and a toothbrush. In his laundry bag were extra socks, his "tin hat," cigarettes, a rifle, and a handgun. He and three pals, all fellow pilots without planes to fly—Robert Bjoring, Leif Kloster, and George Davis—were together when they heard about the surrender. Bjoring and three other men Ty didn't name decided they would head to Luzon's east coast, then go south to attempt an escape to Australia. "There was a feeling that is indescribable hanging around us," Ty wrote. "I was asked by some officers to accompany them to the mountains to try to escape to the Southern Islands until we had really surrendered. I felt it an act of desertion." Instead, he, Kloster, and Davis followed orders, as they had been taught as cadets. As the son of a World War I veteran, and nephew to two others, it's not a surprise that Ty believed he had to obey commands.

"They had just received a call saying for us to prepare to evacuate camp as they were going to have to destroy the ammunition dump below us in a few hours," Ty wrote. Their car couldn't move away from the ammo dump for hours because the road to the headquarters of the 24th Pursuit Group was packed with retreating military units and civilians. "If the Japs had wanted to commit mass murder, they should have known about that jammed road." He and the others got to the headquarters at 3 a.m. and fell asleep on the ground.

Few thought General Douglas MacArthur, portrayed in his own press releases as a heroic action figure, could allow his army to go down in defeat. But MacArthur was in Australia, not to return to the Philippines for nearly three years, and his troops were about to descend into hell.[116]

The morning of April 10, Ty ate some oatmeal, then destroyed things to prevent them from falling into the enemy's hands, including his 12,500-word diary. He recalled the day's events, and that it was his parents' wedding anniversary, in a new diary he began a month later: "We didn't know whether they would accept surrender. Explosions were still going off. We could hear machine gun fire, bombing, low-flying planes. Boy, it is the most nerve-wracking day I ever spent." With little information, "It really put the fear of the devil in all of us. A number of boys broke down under the strain." By nightfall, they still had not seen a Japanese soldier, so they went to sleep.[117]

As Ty ate a breakfast of rice with pea and tomato soup on April 11, Japanese soldiers entered their camp. Initially, they "were really nice to us," trading their cigarettes for American brands. One, whom Ty called "the looter," wanted a camera. Others seized or destroyed food that soldiers were saving for later. Ty and the others turned in their weapons. The next day he mixed powdered milk with water and ate a dill pickle. Soldiers who still had Filipino pesos could sometimes buy food from local residents, and Ty's friend George Davis bought a can of corned beef hash, which he shared with Ty and three others.

Oliver "Red" Allen, an airplane mechanic in the Seventh Materiel Squadron, 19th Bomb Group, was stationed at Clark Field. Like Ty, at the surrender, he put his gun, ammunition, and hand grenades into a mountain of items forbidden to prisoners. "My razor blades went into the pile, as did the can of meat and beans. I guess that hurt the worst; I was terribly hungry. I learned an important lesson: if I got my hand on any food, not to save it. Eat it before it was taken away from me."[118]

Ty said there were air raids but no information about the surrender terms. The soldiers were marched to the side of the road, where they sat on a hill and got a lecture. Ty didn't report whether the lecture was in English or Japanese, but Private First Class Cletis Overton, a member of the 27th Bomb Group, said in an oral history that Japanese soldiers were barking orders no one understood. "It didn't take much for me to realize that they would just as soon kill you as look at you."[119]

Newspapers in Nebraska and elsewhere, some of which Hans and Shy would have read, ran banner headlines about the surrender, although there were few details. The *Evening State*

Journal, published in Lincoln on April 9, could fit only three words into its huge headline: "Jap Bataan Victor." It included a map and a photograph of Ty's friend Wally Churchill on the front page to accompany a story about the last letters received by soldiers' local relatives. Churchill's aunt and uncle believed he had been taken south, as Ty's parents hoped for their son.

The next day, the *Lincoln Star* reported, "Bataan Falls to Overwhelming Japanese Horde: Gallant but Completely Exhausted American Defenders Finally Overcome." Other front-page stories told of the three-month battle, food shortages, and disease. A headline on another story said "36,853 Defenders Face Death or Capture," half of the number of soldiers who were surrendered. There was a photograph of Wainwright, a map of Bataan, and a drawing depicting the tunnels on Corregidor. The paper quoted Stimson, the war secretary: "This is only a temporary loss. We shall not stop until we drive out the invaders from the islands."

The *Kearney Hub* fit four stories across its front page on April 10, including one about the possibility Corregidor could hold out for some time and a three-line headline about sea battles above a map of Manila with a small headline, "Valiant Defenders Lose Bataan." The *Grand Island Daily Independent* paired a front-page story about a German loss to Russia with a map of Bataan and a subhead reading: "Resistance on Bataan Crushed by Yellow Men."

Knowing his troops were too weak or sick to march to a prisoner-of-war camp, King had trucks and gasoline in reserve. A humane march would have been ten miles per day with food, water, and shelter. What happened was the exact opposite. At 4 p.m. on April 11, Ty and the others at the 24th Pursuit Group headquarters were ordered to walk. This was the beginning of what came to be known as the Bataan Death March. Not until January 1944 did Americans at home learn about the march, months after a daring escape by a group of POWs led by

Lieutenant Colonel William E. Dyess, one of the few successful escapes from the Philippines, when their story was published in the *Chicago Tribune* and a hundred other newspapers.[120]

U.S. and Filipino soldiers were loosely organized in groups of about one thousand as they marched nearly seventy miles over the next three weeks—each group marched for a week or ten days—with little food or water. No one knows the exact death toll among American and Filipino soldiers on the march. Some dropped dead of thirst, hunger, or disease. Japanese guards killed soldiers with bayonets if they fell behind or slipped out of formation to drink water from fetid pools or one of many artesian wells. Bodies were left at the side of the road or buried in shallow graves. Friends who kept track of deaths often died later, meaning the military never learned the fates of many soldiers. Journalist Hampton Sides wrote that the best estimate is 750 U.S. soldiers and 5,000 Filipino soldiers died before reaching the prisoner-of-war camp. Historian John Toland wrote that between 7,000 and 10,000 POWs died on the march, 2,300 of them Americans. Historian Frank A. Blazich Jr. wrote that the march claimed the lives of 500 to 600 Americans and 10,000 Filipinos.[121]

Each time Ty and the others met a new group of guards, more of their possessions were looted. Ty was forced to give up his pack, his helmet, and his blanket. He tied his remaining possessions in a towel: a change of uniform, the water purifier, quinine (to treat malaria), a spoon, and his all-important canteen. Japanese soldiers took about fifty canteens from prisoners one day, including David Hardee's. Not having a canteen was a death sentence. Hardee was able to buy one for three pesos from a Filipino soldier who had two canteens. "This was the best investment I made while a prisoner of war," he wrote. Ty wrapped a handkerchief around his head to protect it from the sun. Later, an English-speaking Japanese guard Ty called a smart aleck took his playing cards, pocket watch, and cigarettes. His ring, likely a high school or college fraternity ring, and compass remained hidden in his leggings.[122]

Dyess, a heroic P-40 bomber pilot, included more vivid descriptions of violence on the march than Ty wrote about in his diary. Perhaps Ty didn't see as much violence. More likely, he didn't want to write about such violence in letters meant for his parents. "The wanton murder, by beheadal [sic; beheading], of an American Army captain as the march was getting underway symbolized the horrors that were to come," Dyess wrote about a death he witnessed April 10. "Our Jap guards now threw off all restraint. They beat and slugged prisoners, robbing them of watches, fountain pens, money, and toilet articles. Now, as never before, I wanted to kill Japs for the pleasure of it." Dyess described soldiers being beheaded and sick soldiers being stabbed. "Laughing and yelling Jap soldiers" riding in trucks adjacent to the marching POWs reached out to strike them with rifle butts. "I saw Jap soldiers roll unconscious American and Filipino prisoners of war into the path of the Japanese army trucks which ran over them."[123]

Many on the march reported similar atrocities, including Captain John Coleman, commander of the 27th Materiel Squadron. "I stepped on what looked like a piece of khaki cloth. I slipped back as if in mud. It was human flesh that we were walking on. We could smell decayed human flesh all along the road, where men had been killed, their bodies left where they had fallen. The columns of tanks, truck, and cavalry horses had run over them, pulverizing their bones into pulp." Another soldier saw a Japanese guard throw a POW down on a cobblestone road, and a column of ten tanks ran over him. "When the last tank left, there was no way you could tell there'd ever been a man there. But his uniform was embedded in the cobblestone. The man disappeared, but his uniform ... had become part of the ground."[124]

Japan continued to fight, firing shells south over the marching POWs toward Corregidor, a rocky island of just over two square miles in Manila Bay. Wainwright, the greater Philippines commander, and several thousand soldiers hoped to hold out in Corregidor's vast, but cramped, tunnels that most thought were impenetrable. Japan could not fully control Luzon until it took Corregidor, with artillery that faced both Bataan and Manila Bay. About two thousand soldiers escaped from Bataan in April by swimming or commandeering small boats to get to Corregidor, three miles off the coast, where they hoped for food and protection. The day before the surrender, three thousand soldier patients and eighty-eight Army and Navy nurses at two Bataan hospitals were evacuated to Corregidor. Wainwright held off firing back at Japanese forces for a week, for fear of hitting his captured soldiers. But he couldn't wait forever, and some POWs were killed by their own army.[125]

Winston Groom wrote that some on Corregidor assumed soldiers on Bataan would have a chance to rest and get medical attention. But soldiers were forced to march on roads where powdery dust filled the hot air. "They became the victims of the most studied cruelty that Americans and Filipinos alike had ever or would ever have to endure. ... Because this monstrosity has no dearth of firsthand accounts, no doubt exists that the ordeals these soldiers experienced were beyond embellishment."[126]

In Japanese military custom, suicide was regarded as noble. Being taken prisoner was not. The guards regarded their prisoners as cowards who had killed their comrades. For a thousand years, Japan's old military code, called "bushido," had stood for compassion, kindness, and consideration for one's enemies. But the way Japan and other Asian countries were treated by Western countries in the 19th and 20th centuries changed what bushido stood for. Now it meant "compassion for the defeated was forbidden. Death in warfare was seen to be lighter than a feather, likened to the fall of cherry blossoms. Victory would be sought by any means necessary, honorable or not," author Gregory F. Michno wrote. By World War II, bushido had become the code for all Japanese citizens.[127]

Homma had called for a humane march with food and water, and he could have taken advantage of King's trucks. There was plenty of water along the route from streams and artesian wells. However, Japanese guards, who constantly yelled "speedo" to the POWs, weren't eager to stop the columns to allow them to drink. If a POW or a group of them got out of line at a pond, they were sometimes shot after drinking and filling their canteens. Homma, who claimed ignorance that his orders were not followed, was tried and executed for war crimes in 1946.[128]

Japan didn't yet have enough food on the island for its own soldiers, let alone the twenty-five thousand healthy POWs it expected. Japan assumed the captured troops would have their own food, despite it being taken from them by Japanese soldiers. Tens of thousands of malnourished and sick POWs was a crisis.[129]

After his breakfast of soup, Ty had no food from his captors for days. The POWs ate what they could barter for—often with cigarettes—buy with their remaining pesos, or steal. Guards didn't take Philippines pesos from most soldiers. A few Japanese soldiers traded their yen for pesos; some American soldiers with yen in their pockets were killed on the assumption they had looted the body of a dead Japanese soldier.

Ty bought rice and salmon on April 13 but wouldn't be able to cook it until the next day. He and the other POWs marched into the night, and someone stole part of the rice while Ty slept. He and his buddies acquired more rice, flour, a can of stew, and one bite each of bacon. As soldiers were marched through small villages, residents sometimes threw food to them —"rice cakes, animal sugar cakes, small pieces of fried chicken, and pieces of sugar cane." Then, according to Lester Tenney, a member of a tank battalion and a march survivor, "Suddenly, we heard shots ... Japanese soldiers were shooting at them for offering food to us prisoners." Guards made the soldiers witness the carnage. "Watching this made me feel woozy. I almost started to vomit, but there was nothing in my stomach to come up." Dyess wrote that in one town, after Filipinos threw food to prisoners, the guards "went into a frenzy" killing civilians and stomping on the food. At other times, guards were more lenient. Hardee wrote of one guard who gave him a tin of food and another who allowed him and others to buy fruit and rice on a day when men were collapsing from heat exhaustion.[130]

Soldiers figured out how to march in the middle of a row so they were less likely to be attacked by guards or by Japanese soldiers driving by in the other direction. There were "shootings, beheadings, throat cuttings, stabbings, disembowelments, and fatal beatings," Bill Sloan wrote. Wounded men were buried alive in shallow graves. Helping a straggler could mean death. Sloan added, "As the less aggressive Japanese witnessed examples set by their angrier, more brutish comrades, however, violence and bloodshed became routine. On the Death March, they would escalate into a homicidal frenzy." Soldiers who couldn't understand orders in Japanese were sometimes beaten.[131]

Three days after starting the march, Ty's group reached the area near Lamao, where they were allowed to bathe in Manila Bay. But they quickly moved on because Wainwright was now shelling Bataan from Corregidor. They marched to Orion, where Ty's squadron had been based for a while, then started for Balanga.

The next day, things began to change. Initially, the soldiers had marched under command of their own officers and were occasionally stopped and searched by Japanese guards. Later, they were put into more organized groups led by Japanese soldiers. "It was the first time we had marched under guard," Ty wrote. "Before we were told to march to a place and we went under our own officers. But now, the Japs were taking over, but still hadn't given us an ounce of food. In fact, they had been taking it away from many of the boys who had brought emergency rations. We were really worn out that night after walking and sitting in the baking sun." During the Philippines' hottest month, this practice of forcing soldiers to sit in the sun, even when a shady spot was nearby, became known as the "sun treatment." Soldiers sometimes passed out, then were forced to get up to march again or be killed.[132]

Ty was always meticulous about remembering dates, but by about April 16, confusion set in, which can be a symptom of dehydration. "Things from now on are hard to remember. I was half sick and as a result only half there. We had sat in the sun for the better part of the day though, and the boys were passing out right and left." Others on the march reported the same sense of confusion. Dyess wrote that the POWs were "like Zombies—the walking dead of the Caribbean."[133]

The next morning, near Orani, forty-three miles north of Mariveles, Ty nearly became a casualty. "When I came to where they were keeping the boys, a guard came at me with his bayonet, but hit my towel instead of me. They were beginning to bayonet and shoot stragglers even if they were sick. Boy, you just had to keep going and that was all there was to it." They left early, marching through "towns that were wrecked and the bare concrete and galvanized iron with no greenness anywhere made them hot as the dickens. A number were shot that day as they couldn't keep up. However, at an artesian well, I slipped out of sight into a house, as I couldn't go any farther. I took off most of my clothes and rested. When night came, I went out and took a bath and then came in and slept for the rest of the night." He didn't get up until a Japanese soldier found him and hit him with a stick. He was lucky again, because prisoners who fell behind were often killed where they were lying.

The next day, Ty and a larger group of stragglers started for Lubao, nearly fifty-two miles from Mariveles. Before reaching the town, they were corralled in a field surrounded by barbed wire, "a new place where they had decided to hold stragglers who had slipped out. I guess they figured there were too many to shoot. The road was still littered with dead from months back—anyway weeks." Others also reported seeing bodies along the road and the stench.[134]

That day, they were fed for the first time, half a canteen cup of cooked rice. Ty could barely get it down. Someone, perhaps a guard or another POW, "gave me a handful of a kind of sweet oyster cracker, which I ate O.K. I just couldn't stomach that plain old rice. It was burnt somewhat too." They stayed at the camp the rest of the day and overnight and arrived at Lubao about 11 a.m. on April 20. "I was all in though and went and lay down in the shade and stayed there all day." He bought a piece of brown sugar for two pesos and ate part of it. "I saw [Leif] Kloster for the last time there. He had been there a couple of days. George [Davis] was there, too, but I didn't see him." Some POWs reported that having friends to look out for each other helped them to survive, but Ty and his friends lost track of each other once again. "Well, I didn't know whether I was going to be able to make it or not, but they said those who were sick and stayed behind might be shot, so I pulled myself together. Got some rice they gave out, put some of my sugar and water on it and forced it down. They sat us in the sun on the road for an hour or so. We then moved out for San Fernando."

The capital of the Pampanga Province, San Fernando is about sixty miles north of where the march started. It was a crucial juncture. The Japanese were about to crowd POWs into small boxcars, where many would die of illness, starvation, or suffocation. Some POWs, including Ty, heard rumors about the boxcars and sensed that San Fernando was their last, best chance to escape. Those who survived the three-hour, twenty-mile railroad ride were assigned to groups and counted. Once they were in groups and fenced in at Camp O'Donnell, escape would be nearly impossible.[135]

As his group neared San Fernando, Ty also saw more violence. Others reported more bodies. "After two had been shot that morning—one of whom I saw get shot," Ty wrote, "I knew it wouldn't be long before they were aiming at me, so I looked for a chance to escape, planning to join a column that would come by later." Ty hid in a storage building where, to his delight, ten watermelons sat on a shelf. "I smashed one on the ground and started in."[136]

When the sympathetic owner found Ty eating the melons, he took him upriver in a dugout canoe to his house. "I then gave him ten pesos and the food started to roll in—a can of sardines, sugar, limes, bananas, mangos, canned milk and eggs. I then slept some and ate again that night." That ended his plan to rejoin the march.

Ty was taken to another man, who was with two Americans, Bill and Tom, and a "Mexican" (more likely a Filipino mestizo). Bill Cote, a sergeant from San Francisco in the 20th Pursuit Squadron and the son of an Army veteran, was stationed in the Philippines in 1940. Tom Ward, from Detroit, a private first class in the Quartermaster Corps, had been deployed for about a year. "All of them were four shades to the wind. I gave them some quinine and then got to sleep. We had some crab to eat along with our fruit. We stayed in the dugout all day and that night I took a bath and went to sleep in the house." Ty and Bill "sluffed this Mexican off and were trying to make other plans." Ty never mentioned a reason for wanting to exclude the man, but it could have been racism or that they didn't want someone with them who wasn't in the U.S. military.

They had fruit and coffee the next morning, and Ty took a walk. From April 22 to 27, they debated what to do next. The men who had been taking care of them, Alipio and Angelo, introduced them to Oming, who brought milk and cigarettes, and said they should go with him. They ate a meal of chicken, Ty shaved, and someone cut his hair. By then Ty was out of money, and he felt guilty that the family was feeding them. Alipio suggested that they go to Manila and try to blend in with soldiers who were still there. Even though the United States had declared the capital an open city, it could have meant imprisonment or death for the three Americans.

On April 27, a group of fifth columnists, Filipinos who sympathized with Japan, came to Angelo's mother's house. "Angelo then knew we had to move," Ty wrote. After dark they got into a banca (an outrigger canoe) and arrived at "Oming's on the big river. He was on the bank to meet us and fixed us up food at once—pork and beans, soup." Ty didn't name the river, but the lower Pampanga Province, where he was, is crisscrossed by rivers and streams just north of Manila Bay. Oming's family likely lived in Betis, a rice-farming area then just outside Guagua, and now incorporated into the city. It is about fifty miles north of Manila, around the bend in the road that goes down the eastern side of the Bataan Peninsula. They began to learn more about the family, and Ty realized his dates for recent events might be off a day or two.

"We must have eaten five or six times that day," Ty wrote later about his first day at Oming's family home, April 28. "And they sure treated us fine." Tom was sick, and a doctor examined all three of them. Ty took a walk, swam in the river, took a bath, and helped Tom bathe. "We really ate those days. Oh Yes! I baked some biscuits and some cup cakes, which turned out pretty good." He made more cupcakes and "some small banana pies, which were very dry. I didn't know how to make a cream to put the bananas in. I tried to thicken it with flour and it got lumpy. Then I tried to make a meringue to put on top and it wouldn't beat up for me. I blamed on the climate." He eventually put a crust on top of his creation, and concluded, "I am some pastry man!"

The family decided the three Americans should stay away from the main house. They bought a two-room bamboo hut and set it up about a half mile from their home. Someone brought food three times a day, sometimes by banca. They set up mosquito nets, which Ty declared "were very necessary."

For the first time in months, they ate and slept as much as they wanted. Once word reached them that all the POW marchers and their Japanese guards were gone from the area, Ty and the others relaxed a little. They were introduced to Filipino food and culture, as the family prepared a feast to celebrate a baby's baptism. "They were killing chickens and a pig, preparing fish, and everything else. ... They cooked way into the night." As those preparations were underway, a carpenter was finishing work on their bamboo hut. "The big day—the fiesta. Boy! They butchered another hog that morning and roasted it over an open fire. It sure looked good. However, here is the catch. Bill and I had been eating so much that we had completely lost our appetites and could hardly eat a thing." They saved a fried chicken for the next day. "The whole countryside was here that day and we have been worrying about fifth columnists ever since." Ty's concerns were well-founded. Japanese officials knew some American and Filipino soldiers had not surrendered. "They were to be considered deserters, subject to immediate execution," author Chris Schaefer wrote. Any captured deserter was expected to lead Japanese soldiers to other deserters.[137]

Soldiers in World War II often used canteens that had been made during World War I. The canteen cup—for eating rice or soup—could be attached to the bottom of the canteen. Canteen: E.A Riedel Collection, accession # 2000.598.064 Canteen cup: Helen Tatsch Collection, accession # 1985.536.064 The National Museum of the Pacific War, Fredericksburg, Texas

The canteen and cup were carried in fabric pouches that attached to a soldier's belt. Canteen with attached cup and pouch: Shelley Hatfield Collection, accession # 2003.650.001 The National Museum of the Pacific War, Fredericksburg, Texas

Chapter Eight

Safe and Grateful in a Bamboo Hut

"Boy, these people really do everything for us. ... You couldn't ask for anything more."

—Ty Kokjer, writing about the family who hid him, May 19, 1942

Ty spent a lot of his time in hiding thinking about his future and what he would do when the war was over. Maybe he would build a hotel in a town not too far from Hyannis or take over his parents' hotel so they could retire. Maybe he would be a career Army pilot, or maybe he would fly in the soon-to-boom commercial airline industry. He might go back to college to finish his degree. He wanted to find Mary Tice, whom he had dated briefly, to see if they were still attracted to each other. If things didn't work out with her, he would find someone else to marry and have a family, modeled on those he admired—his parents and other relatives.

Ty wrote on May 10 that it had been a month since the surrender of Bataan. It was a hot, sultry day, with thunder-heads hanging over the nearby mountains. He often reread the two letters from his parents he kept in his pocket. Looking at Ty's diary and his parents' letters chronologically shows they had hope for the future, even when there didn't seem to be any reason for it.

Hans and Shy hoped Ty would write if he had time. He had no way to let them know where he was or what he was doing and that, for the moment, he was safe. They wrote that they were always thinking about each other—what this day was like, what the other might be doing, what they were eating, and about friends and family. Their thoughts were sometimes parallel, especially around holidays or birthdays. In a letter on May 17, Shy wrote that she wished Ty could drop in as a surprise as he had done during flight training in 1940 when he came to Hyannis for Thanksgiving. The next day, Ty wrote in his diary, "I sure would like to walk in and surprise you like I have some times." He wondered which Sunday was Mother's Day. On a lazy afternoon a few days later, he spent an idle hour writing: "It's a bit like talking to you to sit here and think about things in the Hotel Hyannis, the States, well, all over the States for that matter."

He deeply appreciated his situation. "Boy, these people really do everything for us. Not only do they prepare our food, give us a place to sleep, but also wash, iron and mend our clothes, furnish us with clothes we don't have, and give us medical care. You couldn't ask for anything more." Despite his luck, Ty alternated from loneliness to faith in a better future. He missed being around people of his "own kind," but predicted, "It seems that I am destined to get through the war OK. (Just knocked on wood)!"[138]

Aside from knowing that the family members were rice farmers and fish pond owners in Pampanga near the towns of Guagua and Macabebe, all Ty wrote were their first names or Americanized nicknames. Ty left the family surname out of his diary to protect them, should his writing fall into enemy hands. His hosts were four English-speaking brothers: Oming, Doming, Henry, and Bill. They had two older brothers, one a doctor, and a sister, whose names he never mentioned. The brothers' sons also helped, but the Americans found their names hard to pronounce and called them Jack, Pete, Jim, and Dick. Three young women who lived next door—Lulu, 34, Connie, 17, and Gloria, 13—spoke English, cooked, and ran errands. (In one transcript the name is spelled Lri Lu, but Lulu would have been a common name in the Philippines.) Ty referred to only two neighbor families with a surname. Other names pop up without any explanation of who they were. The most frequent is the family's main farmer, Amando, who was likely one of the unnamed brothers or the brother he initially called Henry or Bill.

Ty and his roommates didn't always get along, and after a few weeks, a fourth man was added to the tiny house. Bill and Tom often argued, and the new guy, Pete, "is very dirty," which Ty attributed to his Greek origin, a stereotype Americans often applied to those of Mediterranean descent. Within a few days, Pete, who was in the 31st Infantry, was no longer staying with them. But if Ty wanted company, the little group was about all he had. Sometimes, he spent more time with Bill, other times with Tom.

The house where Ty, Bill, and Tom lived was built with bamboo and nipa palms and sat on stilts. It had four windows. The main room had a ceiling eight or nine feet high, and it measured about six by twelve feet. The second room measured the same, but the ceiling slanted down, as did the porch ceiling. When the tide was in, about half of the house was over water. In the beginning, there was only one cot, and the three men took turns sleeping on the floor, where they were pestered by ants. By July, there were two Army cots and a bamboo bed. They were careful to use mosquito netting to ward off malaria. Four chairs, three water jugs, a table, and firepots completed the hut's furnishings. Water had to be hauled in for drinking and bathing. Their bathtub was a five-gallon bucket. Ty didn't like the taste of the water, and when he didn't bathe often enough, he got rashes, in addition to stinking. Their lamps ran on coconut oil with a piece of cloth for the wick. He compared it to living one hundred years earlier, except they had toilet paper. He was, nevertheless, glad his parents weren't there. He was in a place where most people couldn't understand him, where "you have around 100,000 guys" who wanted you "bumped off," and where he had to "live in more filth and with more insects than are in the whole town of Hyannis."

Their bamboo hut resembled homes of the poorest Filipinos. Ty observed that the people he met were "all classed up." The majority, the hardworking farmers, spoke little English. "They have superstitions and believe anything they hear. They are very easily worked up and one has to treat them very much as a child to make them understand such that you don't hurt their feelings. ... Most of them are happy, though." Their homes had no furniture. At night, they rolled out sleeping mats. They cooked in earthen pots or a metal frying pan over an open fire, and they roasted meat or fish on sticks. Family members each had a bowl to eat out of and used their fingers in place of utensils. They might own a few chickens or pigs that they could sell, donating some of the proceeds to the Catholic church and using the rest to buy clothing and other necessities. For them to move up, Ty wrote, would require "the education of their children or by someone out of their class securing it for them." The middle class lived in town and were primarily merchants or rice mill owners, and "they have taken on our ways a lot and enjoy a car, a movie and perhaps some nightlife."

"The man that owns the land ... lives in a large house" and has storage for his share of the crop raised by the workers. With servants, "he can really play the lazy man." His house would have furniture, and perhaps a piano, as did the owner of the property where he was staying. "The higher class I look at as the grafter. They really are rolling in money, and don't mind making a show of it or taking advantage of someone else to get it. They are the society and I can't say that I care much for them." He criticized the rich for being "radical" for not helping improve "their people." The country, he decided, "is definitely not ready for self government," adding that the current government was "all graft."

His criticism of the wealthier people and others changed over the months. It was their hospitality and financial aid that enabled him to remain safe, to consider marrying one of their daughters, and to borrow money to buy food. Speaking broadly about local people, he concluded, "I get so mad at some of them and at other times, I think there are no better people on earth." He was fond of both the fish pond owner and the Jose family, his main hosts, both of whom he described as "a higher class people than most of them."

After the surrender of Bataan, Japanese forces began a furious attack on Corregidor. In the last days of April and the first days of May, a few nurses, officers, and reporters were evacuated from Corregidor by small planes or submarines, escaping to Australia just before the island and the rest of the country fell May 6. The day he surrendered, Wainwright wrote: "We are subjected to terrific air and artillery bombardment and it is unreasonable to expect that we can hold out for long. We have done our best, both here and on Bataan, and although beaten we are still unashamed."[139]

The same week, the Battle of the Coral Sea raged. Japan won a numerical victory, but its losses meant it could not invade Australia.

From Hyannis, Shy wrote that they would continue to send letters to Ty, even as their missives were returned weeks later. Chan Cobb, Ty's cousin, returned home for a few days, and there was a reunion with Gramp, Robert (Chan's father and Shy's brother), and the rest of his family. "We are so happy," Shy wrote. "We wish for you Ty—we miss you."[140]

Chan served on a Navy supply ship, making transpacific runs to deliver goods to Honolulu and other ports and ships. Ty's friend and fellow pilot Jim DeWolf, who was driving the night of the car accident, appeared in a photograph on the cover of the May 16 *Saturday Evening Post*, which the magazine turned into a recruiting poster for the Army Air Forces. In the poster, six airmen huddle around a map as three planes fly in formation overhead. "Keep 'Em Flying," it says, then suggests men ages 18 to 26 apply to be aviation cadets.

In his months of hiding, Ty had a lot of time with not much to do. In one diary entry, he said he thought he was bored when he drove a mowing machine on a Nebraska ranch, but that didn't compare to his current situation. He got out some, but he and Tom and Bill had to be careful not to attract the wrong kind of attention, even though many people in the area knew families were hiding soldiers. They never went to nearby towns. Ty was always glad to have something to read. One day it was a stack of *Reader's Digest*s from 1939 to 1941. The magazines reminded him of relatives who had been subscribers, including his maternal grandmother, Augusta "Nanny" Cobb, who died in 1940, and his paternal grandfather, Hans Madsen Kokjer Sr., who died in 1932. An article about flying made him realize what he was missing.[141]

When he wasn't reading or writing, Ty played game after game of solitaire, tracking his wins and losses. Some days, he played five games; other days it was twenty-five or one hundred. In 2,350 games he said he played over eight months, he tracked his wins and losses in dollars. His winnings of $965 were dwarfed by losses of $20,933. In that, he was like Wainwright, who played 8,632 games during three years of imprisonment, winning 6.8 percent of them. Toward the end of his time in hiding, Ty played fewer games of solitaire. Instead, he played chess or bridge with Tom.

Ty was bothered by tropical ulcers, open sores also called jungle rot, which can be caused by poor nutrition or hygiene, both of which he had experienced for months. "These darn tropical ulcers of mine haven't cleared up yet. They don't hurt, but the flies get in them and they are messy." Lulu was in charge of most of the home remedies. A treatment with chlorinated lime burned. Another day, she washed the ulcers with hot guava water and fennel. "I about went through the ceiling," Ty wrote. Inspired by a magazine article, he drank some of his own pus diluted in water. That didn't work either. If it had, he would have written an article for a medical journal, he said. By September, he had forgotten to write about the ulcers because they were finally gone.

A few weeks with plenty of food restored Ty's general health, and he regained some weight. He estimated he weighed 130 pounds on April 9 and now was 149, still about 20 pounds shy of his prewar weight. Ty wished for a good T-bone steak, but he was pleasantly surprised by most of the food the family provided to its three soldier guests. Breakfast might be rice with milk and sugar, Ovaltine, bananas, and cookies. One "excellent" meal included chicken soup, boiled chicken, crab omelet, vegetables with shrimp, and watermelon, with American cigarettes to top it off. One of his few complaints is an issue to this day. "We had some good shrimp for supper last night. I only wish though that when they serve seafood they would remove the heads. Of course, they eat the heads and all." Another night, it was carabao with Worcestershire sauce, which Ty found "not bad." Ty learned that *busig-nako* means "I am satisfied" in Tagalog (now called Filipino), the most common language in Luzon.[142]

One Sunday afternoon, after they had vegetables with their rice for the first time in a while, he listed all the food he missed and wanted when he got home. A sampling: Eggs, fixed many ways. Bacon, ham, pork chops. Orange, tomato, grapefruit juice.

Morning snack, ice cream after golf in the afternoon, then steak for dinner. And before bed, sandwiches "or a Dagwood with milk." Pie, ice cream, fruit. Waffles, fried oysters, mac and cheese, chow mein. Divinity, fudge, candy bars. Cigarettes, cigar, or a pipe to follow the evening meal. "I probably could think of a couple more pages—peach and pumpkin pie, baked squash, many other fruit juices, some other cakes, numerous sandwiches, but I guess it can wait until I get around to think of them." And he did. After talking the list over with Bill, he added cantaloupe and watermelon, fried chicken, spaghetti, ravioli, and Welsh rarebit.[143]

For POWs on unfamiliar or starvation diets, conversations often turned to food. "The story of a prisoner of war in the tropics is the story of food. Hungry men can think of nothing else," Hardee wrote as he worked at the prison farm in Davao. He missed good breakfasts. Some of his POW friends "in a mild form of insanity" traded recipes. Bob Bjoring's diary lists recipes and restaurants recommended by fellow POWs.

The farm sat near a river and Manila Bay and had a series of locks to control water in the rice fields and fishing ponds. Ty liked to sit on a lock facing east, the direction of home. Bill thought they would all be home by Christmas, but Ty and Tom thought it would be November 1943, in time for Hans's birthday on the 10th. "I wonder where Kloster, Bjoring, and Davis are now? I hope they are O.K. and that we can have our big reunion in Hyannis as we planned when we were on Bataan." Ty would hang Uncle Tom Kokjer's "No paying guests wanted" sign at the hotel and have a party for all of his Army buddies. He had told his friends that his parents didn't approve of heavy drinking, but on this occasion, he figured, "You would be so glad for us all to be back that you would stand for almost any sort of celebration. Bill and Tom are planning on the reunion also. Bill wants beef steak. He says the rest of the boys can have the liquor."

Ty drew plans for a hotel he might build one day in a town near Hyannis and began drawing a map of Nebraska and nearby states, which survived with his diary. On the other side of the paper, he drew a map of Hyannis, including every building on the town's two commercial streets, Main Avenue and South Railroad Street. Later, he began a map of the Philippines. He said he was surprised by how content he could be with no-where to go, a big change from complaints while he was in the hospital a year earlier. A *Reader's Digest* article about how garages cheated their customers reminded him of his Uncle Tom's car repair business and how the Kokjer name meant car owners could be confident of fair treatment in Sidney, in the Nebraska panhandle. Ty assumed, correctly, that selling cars would be a good business after the war.[144]

Whether Ty should join a guerilla group was a question he wrestled with more than once. Toward the end of May, the three soldiers had the first of several visits from guerillas. "They wanted us to go with them, and when we refused, they asked for our machine gun. I would sure like to know where these people get their ideas. We couldn't even get a cap pistol, let alone a machine gun. Things like this keep you guessing. You don't know if they are fifth column or not, but you've got to be nice to all of them. For even the patriotic ones, if you are not nice to them, they will turn you in if they can get anything out of it for themselves."

Several guerilla groups were led by Americans, whose goals were to get information to MacArthur in Australia. Some raised money and smuggled goods and medicines to prisoner-of-war camps. Others preferred to harass Japanese troops or skirmish with them. Some groups had structures, while some men operated solo or formed temporary alliances. Some worked more closely with Filipinos, who formed many effective guerilla groups. All fought Japan, but some Filipino groups were also anti-American. The communist-led Hukbong Bayan Laban sa

Hapon—the People's Army Against the Japanese—called Huk-balahap, or just Huks, was the best known. That group operated near where Ty was hiding, and it's possible some of the guerillas he encountered were Huks. After a July visit by four guerillas, Ty concluded, "They may be O.K., but without any outside help, they can't do much, and until such a time when they can do some real good, I prefer to stay as I am."[145]

In May Ty expressed mild displeasure in his situation. "One thing I can't blame Uncle Sam for being here. I do blame him for the position I am in at present. It should never have been, I think." A few days later, he wrote: "I have been thinking about this war situation and about some of the laxness of many people. So I have started to write an article that I may try and have published after the war." An encounter with guerillas two month later prompted Ty to unleash an uncharacteristic and harsh critique of American leaders and war policy, which may have been part of his planned article:

Many of these fights would not have to be fought if it were not for the selfishness, carelessness, and stupidity of certain individuals. And there are many of us that know it. Yet! These same individuals, some of our politicians, labor leaders, Generals, capitalists and scores of others, are going around with a full wallet, medals on their chests or getting pats on the back. Those who don't know are giving these men hero worship and a great number of them are accepting it with a clear conscience, as they have either had themselves into believing they were right or are just too dumb to recognize their own faults, yet others are squirming as they should be.

They caused much of this fighting to be still going on by: getting us into the war before we were ready, strikes that sent men to their deaths not equipped, giving orders by guesswork instead of figuring the situation out or asking the advice of others, pushing outmoded equipment onto the Army because they could pull a few strings. Naturally, there has to be casualties and the loss of property in a war but with cooperation it can be held to a minimum.

Then to you who are going to be paying taxes for years to come for the purchase of: planes that were never flown but bombed on the ground; weapons that were needlessly left behind and later turned on their former owners, and for food and other equipment that never saw the men it was intended for. This is what you are paying taxes for and what your children will be paying for in years to come.

Ty laid out his own strategy for an Allied victory: an all-out effort against Germany over the summer; once it got cold in Europe, move planes to the Pacific and bomb Japan. Then back to Germany to "finish them off in the summer of 1943, and if that didn't scare Japan off," increase attacks there, "and by the fall of 1943 things should be all over." He admitted it might take a bit longer. "We will see how my theory works out."[146]

Later in July, there was a firefight not far away between Japanese and guerilla soldiers. Ty feared he and other Americans in hiding might be blamed, even if they had no involvement. He preferred to wait for a well-equipped group that could work just ahead of the return of American forces. But he also sympathized. "I'll admit that these men have reasons for being enraged by the Japs from the way they and their families have been treated, but until they have help from outside, I don't see how fighting could improve conditions. It seems it would only make them worse."

After rejecting previous invitations, Ty wrote on July 29 that he had joined a guerilla group, although he remained in hiding with the family and said almost nothing about the group or its leaders. In early September he wrote about being jittery, and "Ingi, the guerilla," presumably one of his contacts, dropped by. "I may be able to help some in the preparation for the defeat of the Japanese. I may be able to help influence their activities some such that it won't be as hard on the civilians as it has been in the past." He wrote that he wouldn't have to hide when there were rumors of Japanese forces nearby, "and last but not least, it will be an adventure. I feel that if anything is going to happen to me, it would happen just as much here as anywhere else. If that something should happen, you know how much I have treasured the life you have given me, but as I said before, I expect to come through this O.K. and be with you again. I am leaving these books [his diary] and a few personal things here, to be safely kept until I return for them or if not, to be sent to you after the war."

He was told about a guerilla meeting in September, but was also warned it was a Japanese trap. "I am pretty leery about this place yet I don't care a lot about living the life of a guerrilla, and until I hear reinforcements have arrived, I couldn't have any heart in being with them. It's six one way; half a dozen the other." One guerilla gave him a .45-caliber six-shooter in mid-September, "and I feel like Buffalo Bill carrying it around." He gave the gun to Bill. "I'll be running too fast to even think about using it if any Japs get as close as a 45 can shoot."

That same month, he began to rethink his decision. He heard rumors that Japan was being bombed, that its cabinet had resigned, that Japan and Germany were feuding, that Japan lost planes in an air battle over Baguio near Lingayen Gulf, and that it was losing territory in China to Allied troops led by MacArthur. "I am looking for something to pop most anytime, mainly because of the Jap troops moving north which I know to be a fact. Then if even some of these rumors are half way true, it shows Japan to be fairly hard pressed, on most of their far-flung frontiers, which is no help to them." He still felt safe, but, "If the guerillas want me to join them in the near future, I think I shall, as I believe the time is ripe."

In November, he met two Americans on their way to join a guerilla group. There were plenty of rumors about approaching

Japanese soldiers intent on catching guerillas, and Ty thought seriously about joining these men before deciding not to. For someone like Ty, a second lieutenant with little military or jungle survival experience—and having the disadvantage of living in the more open region of rice fields, compared to guerillas who hid in the mountains—becoming an active guerilla would have been a big risk. He was part of a "minute" number, perhaps a few hundred POWs, who had escaped from the Death March. One impediment to long-term survival was the inability for Ty and other Americans to blend into the local population. "Any white captive was a prisoner not only in a Japanese camp, but in Asia," Daws wrote. "His skin was a prison uniform he could never take off." Ty recognized that threat in an August diary entry: "I would sure like to be five inches shorter now, about thirty pounds lighter, have brown eyes and a pug nose along with being about ten shades of brown. I could bluff being a deaf/mute and not even bother about knowing the Filipino tongue." Nearly a third of white prisoners died in captivity, "starved to death, worked to death, beaten to death, dead of loathsome epidemic disease that the Japanese would not treat," Daws wrote. The death rate for those who escaped was close to 100 percent. Some escaped at San Fernando, as Ty had, while some made the calculation that being on their own was too risky. Victor Lear and Michael Tussing Jr., both privates in the 27th Bomb Group, concluded they would be too big a burden on any Filipinos who helped them. "I don't think anybody who was rational thought escape was worth it," Lear said in an oral history after the war. He reasoned that they were too far from home to escape the island, they didn't know how to survive in a jungle, and they couldn't speak the language.[147]

Ty and his roommates heard rumors, most of which he discounted, including that Germany had surrendered. He gave that a one-in-thirty chance of being true. Others included that

Hitler had fled to Spain and that Hitler's second-in-command, Hermann Göring, was in charge, which Ty thought might be good news. Rumors in May that Japan was being bombed were likely a delayed report of the Doolittle Raid on April 18, when sixteen American B-25s bombed Tokyo, the first U.S. attack on Japan. Although the raid caused little damage, it shocked Japan's civilians and military leaders, while lifting spirits at home. Repeated rumors that an American rescue convoy would arrive on July 4 were frustrating. "This business of not being able to believe a thing you hear really gets me, but such is war, I guess," Ty wrote.[148]

In September, he began to draw a map of the world. "It kind of buried my hopes of having aid very soon as there is so much territory that has to be regained before they can get here." He realized the fight going on in the Solomon Islands was more than 2,500 miles away. When Mr. Malorie, the owner of the land where Ty was staying, visited on May 31, the war was all they talked about. The same day, Ty heard a Japanese plane, which wasn't uncommon. "Gee! But I would like to see a swarm of American planes."

Sometimes the rumors Ty heard about the war were true, or nearly so. Burma fell to Japan on May 15. Ty learned something about the June 4–7 Battle of Midway just three days after it happened. It was a decisive loss for Japan and is considered a turning point in the war. Ty heard that all German cities except Berlin were in ruins, likely based on news of the May 30 attack on Cologne by a thousand British Royal Air Force planes, which killed hundreds and left thousands of civilians injured or homeless. Later in June, Japan shelled Fort Stevens, at the mouth of the Columbia River in Oregon, causing no damage or casualties. Ty heard that at Camp O'Donnell forty to sixty POWs were dying per day, that they were being fed once a day, and that they were tortured. The mostly accurate report was hard for him to accept. "You hear some wild stories and you can't believe a thing," Ty wrote.[149]

June to October is the rainy season in the Philippines, and a palm shack wasn't the best for keeping the rain out. Doming and a man named Max soon came to fix the leaky roof. Sometimes, Max also cut Ty's hair. The rains also kept Japanese planes from flying as much, at least near them.

"Seven months from today is my birthday," Ty wrote May 23. "I hope I am not in this place to celebrate it. However, it's hard for me to hope being back with our forces by then, so maybe I should hope that everything goes O.K. for the three of us such that we will be here to celebrate it." His wish came true, but not in the way he envisioned.

Six booklets make up the diary Ty kept while he was hiding with a family in the Philippines. He buried all but one of the booklets in a metal box on the rice farm. The police chief who turned Ty in to the Japanese sent the final book to Hans and Shy. Two or three booklets were lost or destroyed, likely because of water damage. Photo by Joseph W. Brown

Ty kept his 1942 diary in school notebooks published by
the Philippines Bureau of Education. He wrote about his
pre-war life in Manila, the war, childhood memories, and
life among Filipinos. Photo by Joseph W. Brown

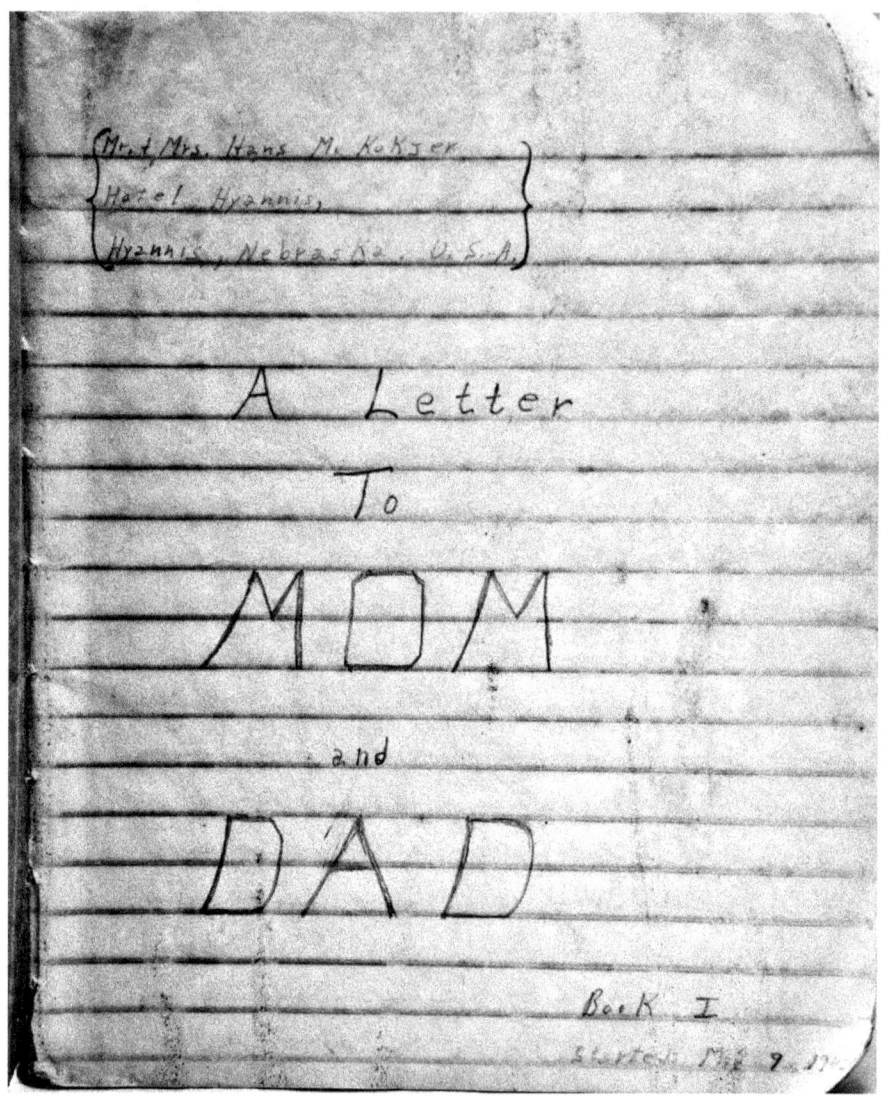

Ty began each book of his diary with his parents' address in the upper left-hand corner of the first page, then printed "A Letter to Mom and Dad." In the lower right-hand corner is the booklet number and the start date: "Book I, May 9, 1942" Photo by Joseph W. Brown

Chapter Nine

Pleasant Evenings and Friends in Pampanga

"I know I will be kidded a lot about the native girls when I get back, so I might as well have some big stories to tell of how all the girls flocked about me over here."

—Ty Kokjer, about being urged to marry a Filipino girl, July 27, 1942

During eight months in hiding, Ty had a unique opportunity to learn about families that were as closely knit as his but also different enough for surprises, criticism, and admiration. "I am learning about the home life and customs of the Filipino which is educational. I will be able to tell you some things about the Philippines when I get back that you won't believe. If paper wasn't so scarce, I would attempt to write some articles. But at present, I am very limited on my paper." Still, he painted word pictures of the families and their lives.

The fish pond owner, who lived nearby, was part Spanish and liked to discuss American culture. Ty wrote to Hans that, if he ever visited, the pond owner could be their fishing guide. Other neighbors were hospitable. Ty liked to visit Ricardo Velize and his brother, Raymon. They didn't speak English, but their daughters did. They played cards, and Ty learned some Spanish. Raymon was mestizo, Ty wrote, making his daughter almost white. In practically the same breath, he recalled it was five years since he had met Mary Tice. Saturday, Wednesday, and Thursday were courting nights in Pampanga, and one evening a neighbor boy brought his guitar and some friends to court Gloria. The group egged Ty on about the Velize girl. "Maybe I should. Her father has plenty of money. She is as nice as most American girls you meet and has been brought up well and is well educated. I doubt if anyone in the States would ever guess she is part Filipino." He described her as tall and resembling a family friend. "I think outside her personality, that hair got me. It is black and she didn't have it done up. It hung about three inches below her waist. If I were to spend the rest of my life here, it would be something to work on, but I know it would never do in the states." When Ty wrote about Mary, he made no effort to explain his attraction to her. But when he considered this attractive Filipina, he felt the need to explain her otherness and give his excuses for being attracted to someone he considered to be of another race. Nonetheless, he decided to visit the Velizes more often. "I did enjoy myself more than at any time since I left Manila." (Apparently Ty waited too long; he wrote in September she had married a lieutenant in the Philippine Air Corps.)

Amando's attractive sister, who was 20, was described as a good cook who knew how to work. "That should be just the girl for me, don't you think? I know I will be kidded a lot about the native girls when I get back, so I might as well have some big stories to tell of how all the girls flocked about me over here." A few days after the first serenade, five boys came to court Gloria. She invited them all in, and Ty joined in the music, singing "Home on the Range" and other Western tunes.

One night in October, he attended a party with Bill following a wedding. He enjoyed the "fine feast" of roasted pig, chicken, other meats, omelets, papaya salad, milk, and bananas. After the party, he accidentally sat on and broke the false teeth that had replaced those he lost in the car crash. He never mentioned getting a new set. But he wrote before breaking the plate that, after the war, he would invest $150 "and have a really good job done."

He appreciated the beauty of his surroundings, especially sunsets, although they made him long for a Sandhills sunset. "The other night, I saw a banca pass with a fisherman in the bow holding a light and a spear. It would really be a picture for some artist to paint. ... Just as the sun was going down last night, the clouds, in spots, took on the prettiest blue color with a baby pink blended in here and there. And then the moon is so bright that it almost has the effect on clouds as does the sun, turning the clouds into colors—a red in one spot, and making the thunderheads stand out by shining on them."

He was often touched by his hosts' kindness. On Shy's birthday, July 17, they made ice cream with milk, sugar, and bananas. Lulu made a special meal—chicken soup, fried chicken, pancit (noodles) with bologna and shrimp, fruit, and custard. He wrote that he hoped this would be the only birth-day she would spend not knowing where he was and whether he was healthy. He assumed Hans had bought her roses and a Whitman's Sampler. In the future, he promised, he and Hans would eat the candy so she wouldn't lose her figure. "I wish it were possible for me to withdraw from your life the hurts, sorrows, and discomforts I have caused you many times. I know you gladly forgave them all, whether they were large or small, and that you will continue to do so always, for I know it is foolish to say they will never happen again, for it is only human nature to at times hurt those who are dearest to us."

On a chilly, rainy day, Ty wrapped himself in a sheet made from four sugar sacks, wore two pairs of socks, and considered putting on a shirt. In the fields, Amando and his helpers were planting rice, barefoot, in short pants, with a single cara-bao pulling crude instruments. Once the rice was planted in mid-July, Ty helped to keep the ducks away by hollering at them. When the rice was harvested in November, Ty helped to separate it from the hull. Days later, it was time to plant the next crop, but Amando was sick. Twenty-five neighbors arrived with thirty carabaos to plant for him. Ty came to respect the carabao. "You eat them, drink their milk, use them as beasts of burden, and I think the kids ride them for pleasure." He learned how to fish without a rod and reel late one night. "Three of us caught nine fish. I caught two. You go into the fish pond and beat the sides of the banca causing the fish to leap into the air and some of them land in the boat. Then you catch them before they can jump out again. It's a great sport. I thought they were crazy until I saw how it works."

The family was planning a feast on November 11, which Ty noted was Armistice Day (now Veterans Day), and some suggested that since there was going to be a party, Ty should use the occasion to marry one of the daughters. Maybe he would, he wrote, surmising Mary would probably be married to someone else by the time he got home. "If I did get married it would strengthen US-Filipino relations and abolish race distinctions." He would have to learn the native language, and she would have to learn English. He would stay in the Philippines for the many business opportunities, he wrote.

Despite pleasant evenings and celebrations, Ty wasn't always happy with his surroundings. As with other criticisms, whether it was the conduct of the war or of people around him, he wrote in early July that his parents should be sure to ask him when he got home. "Part of the picture here at present has to be left unpainted," he wrote, noting that he hadn't found what he thought was an ideal life. He said he tried not to cause friction, but he was still unhappy a few days later. He missed his old Army buddies. "I am so tired of being here with no one that I have anything in common with and being such that we are all about ready to blow up over the least little thing."

His sour mood hadn't lifted when he wrote two days later: "I am sure getting fed up with everything around here, and with most everybody. However, there is nothing I can do about it, so I'll do the best I can under the circumstances. I know the way I feel at present doesn't help the situation any but I'm not going to be the goat of someone else's folly. ... It's dark from so many clouds, so I'll say so long."

With too many family members crowded onto the farm and not enough money, there was friction. One night in late July, a knife fight broke out between Lulu and her brother, Ben, and his wife, Phenung. Lulu accused Phenung of not working hard enough. Ty thought Ben was lazy, but credited his wife, who was 16 years old, as working hard but making the older, unmarried Lulu jealous. Ty brought two of the children and Amando's visiting sister back to stay with him, Bill, and Tom until things settled down.

The next morning, Lulu gave him an ulcer treatment, then went to Guagua to be away from Ben and his wife. By October, Lulu was back, helping them as usual. Ty concluded she was in love with him and expected him to marry her. Although he considered other Filipino women as potential mates, she was not one of them. "I even shudder to think of such a thing." She was a nice girl, and well brought up, but she had a temper that flared when she didn't understand that they were teasing her. Uneducated, she was also "about as beautiful as a mud fence." But a few weeks later, even though Lulu thought Ty should be interested in her, she encouraged him to court visiting sisters, one 18, the other 20. Neither spoke English, "so a merry time was had by all." Although he knew men who had married women from other cultures and races, in the end, he decided, he wasn't likely to be one of them.

Ty thought often of his friends at home and in the military. He was separated from those who went through pilot training with him, either at the surrender or during the Death March. "I will probably never [be] able to contact some of them again. I am very much afraid for those over here, unless they are getting better food and treatment than I think they are getting." He listed their names and their hometown addresses, and asked his parents to find their families if he didn't return: Leif Kloster, Eagle Grove, Iowa; Bob Bjoring, Oslo, Minnesota; and George Davis, Portland, Maine. "When we got anything, we all wanted to share with one another." They could remain calm as they shared opinions. "We really lived in swell harmony. Gee, I hope they are O.K. and all get back to the states in good health."

He never felt the same way about Tom and Bill. Tom, a big man, was too dumb to be mad at, or something was wrong with him. He got dizzy, shook, and often had a blank stare. He had malaria and had been stabbed in the back with a bayonet, which just missed his spinal column. Ty thought it might be the sun, although it could have been the aftereffects of malaria, which include confusion and delirium. Or it could have been post-traumatic stress disorder—PTSD. Tom wasn't the only one reacting to months of exposure to combat. "Last night some lightning struck quite close and it really got Bill. He started looking for a bomb shelter. I jumped, too. ... I'll bet there are a lot of fellows whose nerves will be jumpy for a long time after this is over." Yet, it was Tom who wrote to Hans and Shy after the war, hoping to see Ty again, as he recuperated at the Army's Percy Jones Hospital in Battle Creek, Michigan.[150]

In mid-July, Bill surprised everyone when he announced he had a wife, Nina, although Ty wasn't certain they were officially married. "If I don't come out of this one of the most broad-minded, immodest persons, it will surprise me. But nothing will ever break my ideals." Nina was good-natured, and she cleaned and cooked. She soon went to Manila to retrieve some of her possessions, so it's possible her family was one of many that fled the city when the war started for what they perceived as safety in the country. Ty reported she was heavy and not attractive. But their meals now arrived on time and were more likely to be hot. Lulu also cooked sometimes, and sometimes they prepared their own meals. Ty wrote in August that he wasn't happy that everyone was learning to swear from Bill and Nina. "I try to watch the way I talk as I really admire a man who can hold his tongue."

In September, Bill and Tom argued over whether they should move. Bill was in favor of leaving to protect Nina. Ty agreed with Tom that they should stay, though he wasn't sure it would make a difference. Also that month, Amando met two Americans who wanted to stay with them because they weren't getting enough food where they were. Ty didn't think it was a good idea, given their own food situation, but knew it wasn't his choice. The family agreed to take them, despite warnings from the local police chief, but there is no further reference to them. Beyond that strife, Ty felt Bill walked too far afield to fetch water, but if Ty complained it would cause friction. Still, sometimes there were laughs. Tom was dumped into the river when he capsized in a banca and again when he tried to cross atop a carabao. At one point, they had an injured bird they called Beakie as a pet and played with Heidi, the neighbor's dog. By November, things were on an even keel with the three roommates, but Ty still worried.

Japan had taken over the local news media in January, and with little uncensored news, it is unlikely Ty knew about American military actions that summer and fall. On August 13, U.S. planes bombed Formosa. In October, Roosevelt called for drafting men 18 and 19 years old and signed a $9 billion tax bill, the largest in history. The Army announced that eight hundred thousand Americans were in the service overseas. On August 7, in the first major U.S. offensive in the Pacific, Marines, with air support from the Navy, began the invasion of Guadalcanal and the Solomon Islands. Two days later, the

United States suffered one of the country's worst losses—five ships (in addition to one Australian ship)—off Savo Island in the Solomons. Japan's losses were negligible. The Guadalcanal battle ended with an Allied victory February 9, 1943, some weeks after Ty was recaptured. The decisive action in the Solomons, with a U.S. advantage, would not come until March 1944 when Ty was in a POW camp and more cut off from news than when he was with the family.[151]

Among the interests Ty shared with his parents was the love of food. Hans and Shy wrote about some of their meals and cooking for hotel guests. From the time he began pilot training through his captivity, Ty wrote about what he ate. At times, it was about bounty and pleasure. Later, it was about scarcity. Part of his enjoyment, and sometimes repulsion, was discovering Filipino food, which he wrote about in particular detail in May and June. One morning, they had syrup that tasted like sorghum on pancakes, which reminded him of Nanny's gingerbread. A local specialty called "ballouts" (balut), a hard-boiled chicken or duck egg with the embryo in the yolk, was too much. (Online videos show people eating balut, or expressing their revulsion. It looks like a typical hard-boiled egg, except for the gray embryo, and is a popular street food in the Philippines.) Two-day-old chocolate cake with marshmallows was "still wonderful."

His diet may not have been balanced, but he knew it was better than "most of the boys" in POW camps. "Boy, we are really living out of our class." There was carabao milk every morning and at least one egg a day. They got someone to bake American-style bread, the first they had in a long while, and ate some "wonderful sandwiches." Mostly, they ate rice with meat or vegetables, though vegetables were seasonal, and they didn't always have them. "Haven't had anything special to eat. Rice three times a day. I eat it like a veteran now. I guess I have had it every day now since about January 15, 1942, with the exception of the days that I had nothing to eat at all," he wrote in mid-June. Two days later, he came back to the subject. "I still can't get over how much rice we eat—an average of at least four good sized platefuls a meal—sometimes five or six. I wonder if I will ever eat it again when I hit the states or if I will miss it if I don't have it every now and then."

Like many Americans in the Philippines, he wondered how he would readapt to life when he got home. "I don't know what I am going to do about wearing clothes and shoes again when I get back to civilization," Ty wrote. "Here, all I ever wear is a pair of shorts, unless I go somewhere. And in eating, all we have is a spoon and a fork, so if anything needs cutting, we pick it up with our fingers and tear it apart. I eat with both elbows on the table. I won't know a single table manner when I get back." Dyess had similar misgivings. He and his fellow POWs usually ate while squatting outside. "This started talk about the probability we wouldn't know how to act at splendid functions after the war. We wouldn't know how to balance teacups and salad plates on our knees or handle the table silver of civilization."[152]

The food situation became less reliable beginning in August. One day, their 150-pound bag of rice got wet, and they had to spread the rice on the floor to dry. If they had eggs, they ate three a day, but these eggs were smaller than the ones he ate at home. If they had bananas, he ate eighteen a day. Breakfast one day was fried pork, eggs, diced sweet potatoes, green beans, and shrimp. The next morning, Kellogg's Corn Flakes were part of a meal that included bananas, fried rice, and pork. In mid-September, they were low on food—there wasn't any breakfast, but the midday meal was served early. A few days later, with new supplies, they had a new dish. "We had this young rice fixed in another way last night. You pound it and then toast it in a skillet, then putting milk and sugar on it you have nothing but good old American Rice Krispies."

During a three-day cyclone at the start of October, they were nearly out of food, and there was no way to buy more. The roof blew off Doming's house, so his family stayed with Ty and his roommates. There were ten people in the little hut for a few days. Water rose to the second step of the house before it receded. If the storm had lasted longer, it would have washed away all the dikes and fish.

Ty helped to repair the damaged dikes when the storm subsided. Amando got them some eggs and a chicken, and Doming gave them fish and crabs. Ty made a deal with the Jose family to buy three-fourths of a quart of milk each day for twenty-four centavos. Later in the month, he was upset when Nina and Totung spent $16.50 on food and supplies. "I nearly went crazy to myself at their spending so much. I am really a penny pincher. Especially when it comes to buying soy sauce and catsup as I think it is an uncalled for extravagance, but as long as I am smoking, I can't say anything out loud."

Totung, who lived next door, sold crabs and often shopped for Ty, Bill, and Tom. He knew the provincial governor and police chief and brought news back from his trips to Guagua.

Other storm clouds soon arose. Their source of food in Guagua, a member of the Jose family, said he was under investigation for selling contraband, mainly galvanized wire, and could no longer help. Ty wrote several times about a mysterious "good friend," who loaned them money to buy food. Ty got fifty pesos from the friend and a promise of fifty more every fifteen days. Ty wrote him a check for $55, payable by the Bank of Hyannis. It's unlikely the check was ever cashed, but it's interesting to think what the reaction would have been at the bank.[153]

Hans wrote occasionally about keeping track of expenses when they ran the hotel, and Ty did the same when he was in college, in pilot training, and again in Pampanga. He wasn't the only American soldier who kept track of spending, expecting to repay his benefactors after the war. Army Colonel Frank R. Loyd became a guerilla and was helped by several Filipinos. He visited them after the war and repaid his debts.

Among the people Ty mentioned in his diary, there are at least four possible candidates for the "good friend" who lent him money, although it could be someone he never mentioned to protect the friend. The fish pond owner, who had lived in the States, would have had an understanding of the American banking system, and he was a man of means. The only clue Ty left is that he rowed across the fish pond when he went to meet the friend, as he did when he visited the pond owner. The other possibilities are members of the Jose family, the Velize brothers, whom he described as wealthy, and Connie's father, a professional gambler.

There were times when Ty had little to do, or chose to do little. Most of the time, he searched for things to keep his mind occupied. He enjoyed spending time with people who were educated and could speak English. At a neighbor's house one night in late May, they sang and danced. He described a male guest who was deaf and did not speak as a "good fellow" and an ace at charades. "He can really dance. He and one of the girls put on an exhibition for us and it was really fun. Then they wanted me to dance. So I did some of what I remember of my tap dancing lessons and really had a big time over that." At another gathering a few nights later, Ty reenacted his drum-major days. "We all sang again and then I got to do my tap dance again. They got a big kick out of that. ... Then I got a stick of bamboo and treated it like a baton, and although I dropped it three or four times, they thought it was great. I really am stiff today from my dancing and twirling, though." For someone used to activity, although he never worked out the way modern Americans do, he was too confined. One day in early September, he was jittery with rumors about Japanese soldiers nearby and a looming typhoon. "I have just about torn the house down a couple of times last week practicing gymnastics. I guess being kept in so long I have stored up a lot of surplus energy." He got a workout when he chased a chicken, which he caught and ate.

Ty was often bored. He had run out of things to say to Bill, who thought Ty was showing off if he talked about college. Tom, Ty said, never made sense. Too many local people spoke no English. Ty wished he could take a correspondence course, but decided learning Spanish or Tagalog might be more practical. Old magazines were a godsend, although he found many of their prewar predictions were wrong. With just six unread *Reader's Digests* in early June, he paced himself and decided he would subscribe when he got home. He soon got nine new magazines, including copies of *Magazine Digest*, a Canadian publication. After reading a copy of *Photoplay*, a gossipy, movie fan magazine, he said he hoped never to see one again, despite being a movie fan. He read economics, history, aviation fiction, and other novels, which did him a world of good, putting his mind back at home. He even found himself enjoying copies of the *Ladies' Home Journal*. He likely ignored any women's magazines his mother subscribed to when he lived at home. He mentioned two articles he had read, adding, "You may have read them both, Mom." But there was a downside. "There should be a law stopping the publications from printing these tempting photographs of foods," a sentiment shared by other soldiers. Ty longed for the hams, fowls, pies, cakes, and other food pictured in the magazines. He wondered how he managed to stay cool and happy while missing all those things.

In early November, Ty tried to make himself a new pair of pants. He cut the fabric too small and called the end result "a sad sight." He had begun making a chess set one or two pieces at a time when they first arrived, and he decided he wanted a good set when he got home. In the time Ty made a few pieces, the "old man" next door, who was about fifty, made a full set, and Ty played with him. He was Connie's father, and he spoke English, Spanish, and Tagalog. The family was from Guagua but had moved to the farm for safety.

Music was what Ty missed the most, and he whistled, sang, or hummed his favorite tunes. He wanted to see all the movies he had missed. He thought often of money and what it would cost for him to live when he returned home. He assumed many of his friends had been promoted ahead of him and were getting higher pay. He estimated he had lost $40 a month since the surrender, in addition to the extra pay he would have gotten for flying. Even though he never flew in combat, he had learned a lot. "I worked in or with most every other branch of the service during the time I was in Bataan. I get so riled up every now and then when I think of how different things could be right now if everyone didn't have to be the boss and get their fist in for a handout." On the other hand, he said, if he got back okay, "I will feel I just paid that much to keep my life. If I don't get back, I have been cheated all the way around." Ty never knew that he was promoted from second lieutenant to first lieutenant September 24, 1944. The promotion notice was sent to his parents.

He occasionally mentioned he was running short of paper or that he was starting a new notebook. The first notebook began May 9, and the second began 19 days later, on May 28. The second book covered the last three days of May, all of June, and the first three days in July. In the first book, like the others, he wrote about his daily life in Pampanga. But he also recalled his time in Manila before the war started, the chaotic evacuation to Bataan, and the Death March. In later books, he wrote more about the present and childhood memories.

Ty numbered his notebooks using Roman numerals. Six survived the war. At least two did not. We have books that are numbered one, two, three, six, and seven. The final notebook, which he left with the police, is the eighth. Books four and five—assuming they existed and Ty's numbering system was correct—either disappeared or were destroyed by water. Book three began July 3 and ended August 13. On that day, he wrote that from now on the paper wouldn't be good quality. "You may never see this anyway, so if I do lose them, I won't have lost as much in the next book." That entry was followed by a two-week gap, likely because of a missing or destroyed notebook.

Book six began September 1; book seven began September 26 and ended November 18. The final book began December 4 and ended December 17, the day he was captured. In that book, he often skipped days, and many entries were short because he had little time to write or little new to report.

In early October, he read "Little Boy Blue," a sentimental 19th-century poem by Eugene Field about the death of a child, and decided to write one of his own. The six-verse poem begins with this quatrain:

The Parents of Little Boy Blue
Your little boy blue has grown to a man,
His little toy soldiers he has left for you to command,
He is carrying the torch of America's reply
To those in Flanders Field where poppies blow.

It ends with these lines:
The days roll by and the guns roar on,
Blasting the peace of the world,
Your boy is there, but you know not where,
And you wonder and worry and fret.
But you can bet; he's upset.
And, it's three-to-one
He'll be home on the run
When it's done.

Of his effort, he declared: "And so, for great poetry, I quit."[154]

After some guerillas left the area in mid-October, everyone relaxed a bit. "Little Nina," Amando's five-year-old daughter,

learned to say "I love you" in English. One day, Ty played hop-
scotch with Consing, to his regret. On one hop, he lost his
balance and fell into a fish pond, hitting his pelvic bone on the
cement on his way in. He was in pain the next day, but never
mentioned any treatment.

On Saturdays, he wondered about Nebraska football games
and the "quite normal" lives many people back home were living. He said he was tan and could stand to lose five pounds. He
imagined his parents eating their Sunday supper in the hotel
that was full of cattle buyers and hunters. He found pleasure
where he was, and he longed—often without much hope—for
those quite normal days he justifiably feared he would never
see again.

A prisoner at Cabanatuan POW camp stands next to a well and a small garden in 1942. POWs raised okra, eggplant, and pigweed. They lived in thatched-roof barracks. Ty arrived at the camp in February 1943. This photo was released February 10, 1945. Signal Corps-SWPA-45-1199, original negative lot 14289, National Archives and Records Administration

Japanese officer reviewing a group of American officers at Camp No. 1, Cabanatuan, Philippine Islands, in December 1943. This picture is reproduced from a Japanese magazine published in Manila and received by the wife of Lt. Col. Leo C. Paquet.

The photograph in the Red Cross newsletter depicts soldiers at Cabanatuan POW camp and was first published in a Japanese magazine. Taken in December 1943 when Ty was there, it is likely an example of "dressing up" the camp and some soldiers to create the false impression that Japan was observing the Geneva Convention. By late 1943, no POWs had clean, well-pressed uniforms.

FAR ✚ EASTERN
PRISONERS OF WAR BULLETIN

Published by the American Red Cross for the Relatives of American Prisoners of War and Civilian Internees in the Far East

VOL. 1, NO. 1   WASHINGTON, D. C.   AUGUST 1945

American Prisoners in the Far East

By Captain H. L. Pence, USN (Ret'd)

Captain Pence, who recently retired from the Navy, is now in charge of the Far Eastern section of Relief to Prisoners of War. Mr. John Cotton, who formerly held that post, returned to the Philippines in June.—Editor.

One year ago the largest concentration of American prisoners of war in the Far East was in the Philippines. At the same time the largest concentrations of civilian internees were in the Philippines and the Shanghai area. With the liberation of the Philippines in the first half of 1945 many prisoners of war and civilian internees who had been confined in that area were released. Of the remainder, a number were evacuated to camps in Japan and elsewhere in the north, while others are still unaccounted for. At the present time the greatest concentrations of prisoners of war are in Japan and Manchuria, and the greatest concentrations of civilian internees are in Shanghai.

At the present time (July 1945) official lists of American nationals in the Far East show a total of approximately 14,000 known prisoners of war and about 3,400 known civilian internees.

There is some evidence that prisoners of war and civilian internees have been moved several times and to various areas. Reports of these moves have not always been available. It is hoped that more accurate figures can be given in due course. In the areas where camp visits have so far been possible they have averaged

about one per camp per year, but conditions in Japan are now such that the authorities are further curtailing the visits.

A press report from Manila in the latter part of June gave some statistics released by the Recovered Personnel Branch of the Adjutant General's Office of "Army Forces Pacific," the latter being the new title of General MacArthur's command. It was reported that of the more than 18,000 members of the U. S. armed forces captured at Bataan and Corregidor by the Japanese in 1942, 10,000 enlisted men and 2,000 officers had been sent to prison camps in Japan and other northern areas, 3,260 had been officially reported dead, and 1,447 had been rescued. About 1,500 are still missing. As to civilians, 5,844 have been liberated in the Philip-

pines. In general, the Japanese left the weak and sick in the Philippines, the others having been sent to northern areas, presumably to labor camps. As for commissioned personnel, officers of the grade of colonel and above were usually sent to metropolitan Japan or Manchuria. Bilibid Prison, in Manila, was the staging point for the shipment of prisoners of war, and official records were maintained there.

Relief Shipments

Relief so far has been supplied by providing funds, as well as by forwarding shipments of food, medicines, clothing, and other articles whenever it was possible to make arrangements for delivery by the Japanese. Neither method has been entirely successful. Local supplies have been high in

Japanese officer reviewing a group of American officers at Camp No. 1, Cabanatuan, Philippine Islands, in December 1943. This picture is reproduced from a Japanese magazine published in Manila and received by the wife of Lt. Col. Leo C. Paquet.

The American Red Cross was the main source of information for families of soldiers overseas. Hans and Shy kept copies of several newsletters. This one was written in July 1945 and published in August, five months after the Philippines were liberated. The article reports on the progress of finding and repatriating prisoners of war.

Chapter Ten

Dreams of the Past, Hopes for the Future

"Yes, I have had a lot of fun in these 22 and a half years. Except that last half."

—Ty Kokjer, writing about memories of home, June 22, 1942

Sitting in bed one night and smoking, Ty wrote that he had dreamed about having talked to his mother. "I was letting you know I was all right, that I knew you worried about me a lot, but that I thought you had an inward feeling that I was all right just the way I have that feeling about you. ... You had tears in your eyes and I kind of woke up afterward and my eyes were also wet." Another night, he woke "repeating 'Everything is O.K. Mom.'" He imagined Gramp, widowed two years earlier, was in the same boat—alone, two hundred miles from his two adult children, and thousands of miles of water away from Ty, where half of the people around him couldn't understand him.

He often mused about the past and made plans for the future. He calculated he was living in a time zone sixteen hours ahead of his parents and tried to think about what they might be doing at certain hours of his day. He appreciated what his parents had done for him and listed two experiences he would go through again to be home: withdrawing from the university to have his appendix out and breaking his leg in the car accident during pilot training. He said it gave him a thrill to know that his relatives were "wholesome people respected in their communities and with the faith and foundation for good American living." He listed the families of Uncle Tom, Nanny, Uncle Em, Gramp Kokjer, and Uncle Bob. He thought of his cousin Margaret Cobb as a model relative. "The more I think about Margaret, the better I like her and the prouder I am of her. She is one swell all-around girl. If I ever had a daughter and she can do half the things and hold down half the honors Margaret has, I will be more than satisfied." Thinking about his boyhood activities and trips, he imagined a potential son. "I hope that if I ever have a boy of my own, he can have as much fun and do everything there is to do like I have done. Except, of course, getting stranded in the rice fields of lower Pampanga in the Philippines or having to be in a war." He thought about his cousins Chan and Bobby Cobb on their birthdays, and he hoped relatives serving in the war would make it back home. (Chan joined the Navy before the war, and Bobby joined the Signal Corps soon after the war started.) Memories of a band playing the *Star-Spangled Banner* as the flag was lowered at sunset made him emotional.

The only photo of his parents Ty had with him was taken on a happy Minnesota vacation in 1933. When he got home, he wanted a new picture, even, he wrote cheekily, if his mother weighed two hundred pounds. In the letters Ty never received, Shy wrote about her successful weight-loss program. Ty didn't know that Jordan Madsen Kokjer had died. The Congregational minister "was your daddy's last uncle in the U.S.," Shy wrote in her scrapbook, and one more link to family in Denmark gone. In another note she wrote that she had come downstairs that morning in late June to find dirty dishes in the hotel kitchen, which she washed, and reported that the Sibbitt family ranch house had burned down overnight, putting Mr. Sibbitt in the hospital.

In late June 1942, Shy confessed in a letter that they weren't writing often enough, but they would if they were getting letters from him as often as they had when he was in pilot training. "But we will get them, and so will you," she added. In her scrapbook, Shy pasted a picture of his cousin Janet Kokjer, then 11, posing with her baton. In her note, Janet wrote, "Don't worry about Ty too much."

Shy sometimes wrote about visitors at the hotel, including Ty's friend Jack Dinsdale. He and a friend were buying cattle to ship to the Dinsdale family's feed lot in Palmer, Nebraska, near Grand Island, where they fattened cattle raised on Sandhills grass and sent them to the stockyards in Omaha. Ty mentioned Jack several times in his diary. Jack was drafted into the Army in 1942, served until 1946, then returned to the family business.[155]

On a trip in July to attend a wedding in Lincoln, Hans and Shy stopped in Wahoo, Clarks, and Funk to visit family. Hans met with state Supreme Court Judge Frederick Messmore. He and his wife, Jennie, hadn't heard from their son, Hiram, who was in the class ahead of Ty's at Kelly Field. Ty had dinner with the Messmore family when Hi graduated. Judge Messmore believed that Hi and Lieutenant Wally Churchill, one of his son's classmates and a friend of Ty's, had flown to Mindanao, where they hoped Ty was, and were likely safe. The two couples corresponded throughout the war.

Hans, Shy, Tom's wife, Isabel, and Hans's youngest brother, Emerson, used their contacts or wrote to strangers in an attempt to learn something about Ty after they stopped receiving letters from him. They rarely learned anything useful. Hans and Shy received a letter from Colonel Charles Backes, who was evacuated from the Philippines to Australia and returned to the United States. They had written to him after reading a newspaper article he wrote that was printed in a number of newspapers, including the *Lincoln Star*, on June 17, 1942. Backes's response was equivocal. He said he believed all those who landed in Manila on November 20, 1941, including Ty, were sent to Del Monte on Mindanao Island, about nine hundred miles south of Manila, where there was an airbase. Backes was correct that that had been the plan, but the start of the war and the lack of planes ended it for Ty and nearly all of the 27th Bomb Group. Backes also seemed to know that not all the airmen had left Luzon, writing, "In a very short time our air corps troops were considered among the best of infantrymen on the peninsula." Backes knew Lieutenant Messmore, who was healthy when he last saw him, and Lieutenant Churchill and assumed all three—Messmore, Churchill, and Ty—were together in Del Monte, where there wasn't much fighting. "I regret that though the name of Lieut. Kokjer seems strangely familiar to me, I cannot recall meeting him in the Philippines." At worst, he wrote, Ty was a POW, "and when this terrible mess is brought to a successful conclusion, he will be returned to you safe and well."[156]

Like other Americans, Backes as yet had no knowledge of the Death March and the terrible conditions in POW camps. Hans and Shy shared the letter with Judge Messmore, who at that point had not heard from his son. Messmore replied, "I have a feeling that most of the flyers who were on Luzon at Christmastime subsequently got to Mindanao." He said he got information through a circuitous route that he could only say involved a doctor in Australia. He said several pilots, including Churchill, were hiding in the hills with the Moros, a Muslim community in the majority Catholic country, and were well provisioned and cared for. "I am quite sure that Ty is all right and I believe he is on Mindanao."[157]

On July 6, 1942, Hans and Shy received a letter from the Red Cross saying Ty's name did not appear on any casualty list through June 12. "We are trying to procure a mailing address, but Lieutenant Kokjer is probably stationed a great distance and mail was returned pending his arrival. ... If Mrs. Kokjer hears from her son directly may we ask that you let us know immediately?" The Red Cross letter added, "The Japanese Government has indicated its intention of conforming to the terms of the Geneva Convention with respect to the interchange of information regarding prisoners of war." It expected to receive a list of prisoners from Japan. The War Department would consider soldiers missing in action from the date of surrender of Corregidor, May 7, 1942, or twelve months from the last contact. At that time the department would make a final determination whether to assume the soldier was missing or dead. The soldier's pay would continue during that time and if he was confirmed to be a prisoner of war. For Hans and Shy, that letter held both good and bad news. Ty wasn't known to be dead or injured, but it was clinical in describing what could happen within a few months.

By October 1942, Hi Messmore was home on leave. Newspaper stories told of his Purple Heart and Silver Star for flying in the Solomon Islands battle. Hans and Shy saw the story, and she wrote, "We hoped they might say something about you, Ty, but they didn't." Hi was one of the last pilots to get out of Mindanao before Japan took it over. He told his family he was certain Ty had never left Bataan, but had not been injured when he left, news that soon made it to Hans and Shy. Another story about the same time described a mission flown by Ty's Kearney friend, Jim DeWolf. Shy pasted it in her scrapbook.

In the Philippines that June, however, Ty was wishing his unit's planes had arrived so he could have flown away with his other pilot friends. But he didn't feel as if all his time in hiding was being wasted. He appreciated the advantages he had in life and what he had accomplished. He could admire what he called the "simple" life of Filipinos, how good some people were, and how treacherous others could be. He was reading books and magazines he would never have chosen for himself and learning new things. As he thought about his future, ideas for a career and life streamed onto pages and pages of his diary. Ty wrote often about the hotel business, noting in mid-June that they had signed the deal for the Hotel Hyannis two years and eleven months ago. He planned to be home by the fifth anniversary. Chadron, at the junction of two highways in the northwest corner of the state near the South Dakota border, was the focus of his expansion plans. The town—popular with hunters, hikers, and horseback riders—could use a hotel with a banquet room, and he could hire Chadron State College students for many of the jobs.

Shy wrote in June that the balance in Ty's bank account was building up, and when he came home he might have enough money to buy a hotel. Across the globe, Ty, who worried that construction costs for a hotel would be too expensive right after the war, thought about living with his parents instead and leasing land to raise cattle, staying in the Army, or becoming a commercial airline pilot. He wanted to keep learning. English, typing, meat cutting, commercial law, home economics, advertising, political science, and government topped his list. He could return to the University of Nebraska. "Well, when I get home, if it is at the end of 1943, on what I will have, I could buy a car and go to school for two years. But do I want to do that?" By 1944, the new G.I. Bill would have paid for Ty to finish his undergraduate degree. In a sour mood one afternoon, he said he wanted to be cheerful when he got home. "If anything will put me back in shape and give me a younger outlook on life, I believe college is the thing. Marriage would be just as good, maybe, or me taking Dad in golf, or Dad and I taking Mom and anyone else in bridge."[158]

Married or not, he eventually wanted his own home and found the *Ladies' Home Journal* provided him with plenty of ideas about house plans and furnishings. After studying the magazine's monthly house plans, he decided he could design his own. A three-bedroom house with two bathrooms and a basement rec room sounded good. "Yes! I am all set to live. Now all I have to do is find a wife. I have everything else figured out (maybe?)." If he didn't marry, he would build a one-room house with a kitchen on the main floor and a rec room in the basement, where he would sleep. He wanted dark green carpet, a fireplace, a davenport that converted to a bed, a den with bookcases, leather furniture, a heavy work table, and linoleum on the rec room floor. Round corners would make it easier to clean. It was something he doubted most young men would think about, but he surely knew how the hotel was cleaned and probably had done some of that work himself.

For entertainment, he wanted a Ping-Pong table, a radio and phonograph for dancing (which he would buy as soon as he got home), an electric train set, and an outside playground for badminton and croquet. In the winter, he would flood the yard for ice skating, which he thought might be a good idea at the hotel. "Boy, some of my ideas are so brilliant that I don't even waste the paper to write them down." He thought about how much it would cost to raise a family. "If I can't give a child everything I have had, I don't believe I would want any at all." He concluded at one point that he might have been reading too much about home life and was "slightly off my head." But he went right on, thinking about how he would structure his life as a hotelier. His cousin, Bobby Cobb, "a super salesman," might be a good addition to the staff. Ty would take every other Sunday off, plus one afternoon to hunt or fish. He would visit Lincoln and Sidney periodically. He decided that if he lived to be 25, he would live to be 70 and retire at 60 with at least $25,000 in savings, giving him $2,500 a year to live on, which he thought was plenty. If he was at home or in Sidney, he would play golf, and "Aunt Izzy would have one of her good roast dinners" with plenty of cold milk. He had enough to eat now, but said when he thought of food from home, he got hungry.

He thought about travel. He had never been east of the Mississippi River, except for his quick trip to the naval air station at Pensacola, and wanted to see Maine and more of Florida. One of his friends hoped to land a job guarding the Grand Canyon, which intrigued Ty.

Always good with dates, Ty reminisced about his past life. On June 21—the longest day of the year, which wasn't special where he was—he recalled he had met Mary Tice five years ago while working at the Metzger family ranch just after graduating from high school. She was 16 years old and about to be a senior in high school. Ty was on his way to pick up the mail, and he invited her to ride along. They spent the evening together. "That night I didn't go to bed with the chickens, but was trying to impress Mary," he wrote. He recalled a nearly full moon. The next day, he took part in his first branding near the Snake River northwest of Hyannis. It did not go smoothly. A calf spooked his horse, which ran off. He eventually got the calf branded. He drove back with Mary in the truck, which got stuck. A team of horses had to pull them out. His art-student friend, Marvin Metzger, teased Ty for "one-arm" driving with Mary. "Yes, I have had a lot of fun in these 22 and a half years. Except that last half."

He wrote about some of the music he missed, including the 1941 Andrews Sisters hit song "I'll Be with You in Apple Blossom Time." He said it made him think about Mary, "and I wish for the things to happen that the words of song express." The lyrics start with the title and continue, "I'll be with you to change your name to mine."

In June, he wrote to his parents that he might send Mary a proposal as soon as he could get a letter out. "She couldn't do any more than say no or cut off diplomatic relations altogether." He reasoned that it was probably better to wait, as they both would have changed since they last saw each other, although they exchanged letters until he left for Manila. Shy had forwarded one of Ty's letters to Mary's mother to send to her daughter. Ty wanted to see Mary again before deciding to marry someone else. "Don't get the wrong idea that I am just batty on the subject, but by that time, I will be old enough and I do want a home before too late in life." If he arrived home to discover that Mary was no longer interested in him or was married, he admitted he would be disappointed. "But I shall try to go on the theory that 'everything happens for the best.' If I ever get back again, I will sure believe that."[159]

His ideas about marriage seem more forward thinking than the pattern for many postwar couples. It also mirrored his parents' roles as they managed the hotel together, or alone if one of them was away. Ty said his wife should be "one whose joys and sorrows I can also have as well as her having mine. I don't want one who is to keep the house and look after the children. I want a share in that myself just as much as I want her to have a share in my work and knowing how I feel about it. ... I want someone who will be happy when we might be down, and of course, it will make us both happier if we are up. And someone that doesn't feel we have to keep up with the Jones' family."

There are two jumbled pages of text that are part of a letter to Mary in the National Archives diary transcript and another in the second transcript, now at the Library of Congress. They do not appear in the later transcript his parents had made, and it is unlikely the letter was mailed. In the letter Ty laid out a sort of marriage proposal and options for a life they might spend together. He said he heard rumors that many women were working outside the home. "I don't know about your ideas on that," he said. On the assumption Mary had finished her one-year business course by the summer of 1942, she could have been holding down a job as so many American women did during the war. She might have worked at a war production plant; Ty's cousin Phyllis Kokjer made bombs at an ordnance plant in Mead, Nebraska, near her hometown. Or Mary might have become one of the "government girls," typing away in Washington and elsewhere. An estimated two hundred thousand young women worked in offices in the capital, and they crowded into scarce boardinghouses at night. Given Ty knew women were holding down jobs, he probably would not have been surprised to learn that Mary and Phyllis were among them.[160]

"Many girls now like to work even after they are married," Ty added. "I don't know how you feel about it but if so you could help a lot in running the hotel. You could oversee the chambermaids and the waitresses and advise on the meals, etc." Most of the time, Ty wrote to his parents that he envisioned a future in Hyannis or a nearby town, but he wrote in June it might make sense for him and Mary to study hotel management. In his letter to Mary, he wrote about Cornell University in Ithaca, New York, which had a course in hotel management. Afterward, they could work for a large hotel in a big city. "I don't think I would and I doubt whether you would be interested in spending too many years in a place like Hyannis," he wrote. "In time I hope to own two, three or more small hotels. Run right they make as much money as a large one without half the effort, worry, or chance of being completely wiped out." A chart in the Archives version of the diary lists likely costs of living in Chicago, New York City, Colorado, and New Jersey.

Ty thought his chances in commercial aviation after the war might be slim with so many pilots leaving the military. But he missed flying and thought he was in good enough shape to pass a physical that would allow him to fly. "I never knew how much I really liked it until I was deprived of it with the thought that I might be out of it for good. At Oxnard, it was the newness of it and the acrobatics that made me enjoy it so much. At Randolph, I didn't care so much about it, but then at Kelly, the cross-countries, night flying, instrument and formation flights all held fascination, and I really want another taste of it before I leave it for good." On an August day, as he watched children fishing in one of the ponds and Japanese planes flew overhead, he wrote, "Today would be a really swell day to fly. The air is quiet and there are a few lazy clouds floating about the sky. A year ago now I was having some of that good flying." Staying in the Army might make sense, he wrote to Mary. "This is what I would really like to have happen: As soon as the war is over to get married and stay in the army at least six months such as I could make plans with you what to do from there on out. The reason for wanting to stay in the army at least that long is that I want a chance to fly some more, for after I leave the army I don't believe I will ever have as good a ship to fly."

He acknowledged she might not be ready to marry him right away. "Maybe some people could show themselves through letters but I feel that I need to be with you to win your respect and love." He admitted he had faults. But he also said he had changed since they last saw each other, and he felt older. "I am perhaps somewhat spoiled being the only child in the family and have had my own way too much and have gotten to do or have about everything I wanted." But he said he was always thinking about her, especially at night as he fell asleep. "I have always thought more of you than any girl I have ever known. I have wanted to be with you so that you could get to know me better in every way. I feel ... that when the time came for me to get married that you are the one I will want to marry. This is putting it very bluntly, but I am no master of words. I wanted very much to see you before I left but things just ...." The unfinished letter ends there. But there is a sign-off, "With love and thoughts of You."

In the second transcript, at the Library of Congress, there are a few sentences on a page that also contains a diary entry dated August 13 and is the end of Book 3, which Ty said he would bury. He was writing to Mary about food: "I like to try and cook some myself. However, we could eat a lot at the hotel if you wanted to. But we can talk about such things at a later time if things work out. I have wondered how you like to cook. I like good food but I am not hard to please especially after having eaten rice for the last seven months. Plain rice, soft rice, Spanish rice, rice and meatballs, rice croquettes ..." There the transcript ends. The next diary entry is dated September 1 at the beginning of Book 6.

Ty's wish for marriage never came to fruition, and there is no evidence that Mary knew how much Ty cared for her. Like many soldiers, Ty seemed to grow more in love with his hometown sweetheart the longer he was away. Ty received a letter from Mary shortly before he left for Manila, but it did not survive. If he was able to get a letter to her from Manila, there is no record of it.

There were no Fourth of July celebrations where he was, so Ty wrote to his parents about past celebrations. "I just sharpened my pencil, put a pillow on the chair, and lit a cigarette, so now, I am all set to visit with you a while." He had returned from visiting a neighbor who gave him a glass of milk and a pack of cigarettes. "People are always wanting to give us something." The midday meal was also a hit. "What a dinner! Fried chicken, rice, canned tomato soup, (which I used on rice to kill its taste), canned beans fixed with meat, bananas, sweet coconut, and some candy and cookies." He wrote about dances, rodeos, and the start of haying season back home. "I have spent some swell Fourths and have hopes of spending many more." One Independence Day was more of a dud, however. Hans and Shy had given the help the day off in 1940. "Then business flowed in thick and fast. I washed dishes and everything else." Afterward, he went to a dance in town, "which was a flop." July 4 dances in other years were more enjoyable. He recalled holidays as a child in Kearney, Clarks, Funk, and western Nebraska.

He assumed the holiday at home would be good for selling bonds to increase war production. He and Bill agreed it was not a good idea to draft movie actors or big-name ballplayers and assumed most others in the service would agree. "They can do more for the morale of the army out of it than they could do to strengthen the fighting power in it."

Hans also wrote about current and past Fourths. He and Shy had gone to the country so Izzy and Tom's daughter, Ann, then 9 years old, could shoot off firecrackers. "It made me think of the time you shot one off in the car. That didn't get over so good or did it?" In 1942, the United States marked the holiday by sending planes on their first European mission, joining RAF planes to bomb German air bases in Holland, although the mission was largely unsuccessful.[160]

On a day in late September, Ty surmised it was the opening of the fall football season, if there was anyone to play. Nebraska fielded teams, although about two hundred colleges suspended football until the war was over. He pictured all the family's belongings in rooms 11 and 14 at the hotel, in all the places they had lived, and in his grandfather's house in Funk. "The present means nothing. I am living the past over again and wishing for the future. I think of you more than you probably realize." All during late September and October, Ty wrote about what he had been doing a year earlier—finishing flight training, the quick trip home, and the enjoyable time on the trip west to see the sights of San Francisco and visit family before he left for Manila. "A year ago today was the last time I saw you," he wrote October 15. "We had a swell trip together, and were making plans for being together again. We saw you off in Oakland and then missed the ferry back so had to wait around for some time." Had he known he would miss the ferry, he could have spent a few more minutes with his parents before their train left. Now, he was restless. "I would sure like to have something happen as it is getting tiresome here. Yet, I guess I should be happy I have not been a captive all this time and if any of those boys come out with their health, they will really be doing well."[162]

Hans and Shy longed for information and were confident that their son would write when he had time. They sent some letters to Australia, hoping Ty was one of the few pilots who were evacuated from the Philippines. They figured out from news reports it was unlikely he was flying. On October 26, Ty recalled his last telephone conversation with his parents. Shy wrote in her scrapbook that he sounded happy on the call. They were back in Hyannis, and he was about to board the ship. It didn't help his sour mood. "A year ago now I was on my way here but now I am here wishing I was there. ... They know one kind of life and I know another, and then being cooped up here for almost six months doesn't help any." He wished he had been stationed at Hickam Air Force Base, now part of Joint Base Pearl Harbor, although he acknowledged he might have been killed December 7, 1941.[163]

Ty mentioned his plans to bury his diary again on October 18, after several tense days with guerillas attacking Japanese soldiers in a nearby town, then staying on the rice farm. "This morning, I got together all that I have that I want to keep in the way of papers until after the war and got a tin can and buried them so now if I ever have to take off in a hurry, I can come back at a later time and dig them up. I also told one of the Filipinos that if I didn't return after the war was all over, that he is to dig them up and send them to you."

On November 4, Ty wrote about his ship's short stay in Honolulu: "A year ago today was the last time I flew."

He wrote to Hans on his birthday, November 10. The previous year, he had missed his father's birthday altogether, as he crossed the international dateline on the ship, skipping from November 9 to 11. He was not quite sure of his elders' ages, but figured his dad was 54 or 55 and Gramp was 72. (In

November 1942, Hans turned 54 and Gramp turned 70.) Ty looked forward to being with them for many years.

The next-door neighbors made rice cookies for Ty as a treat on Hans's birthday. As he wrote, Ty was smoking a pipe and looking out the window, where the sun was behind a bank of clouds. "I have thought of you every hour in every day, but today especially I have gone back over the good times we have had together and have been hoping for those times to continue in the future." He recalled his dad playing on the floor with him when he was a child, skating, tennis, cribbage, and golf. Hans worked as a traveling salesman for many years, and Ty missed him while he was away, although he accompanied his father a few times, including July 4, 1936. "I think I was somewhere in southeastern Kansas with you, Pop. I know the day Landon was nominated we were in Coffeeville." Kansas Governor Alfred M. "Alf" Landon was nominated in mid-June 1936 to be the Republican presidential candidate. As was then the custom, he was not at the convention and gave his acceptance speech in mid-July. Roosevelt beat Landon with 60 percent of the popular vote and won the Electoral College 523 to 8.[164]

Ty also figured out that his parents sometimes gave up food they liked to provide for him. "I never realized before, and probably don't yet, the sacrifices you and Mother have gone through for me and all of us." He wrote in June, "I'd be willing to start life all anew at the age of zero if I knew you were the ones that were going to be my parents."

In the same week Ty wrote about marrying a Filipino woman and staying on to run a business. He added, "I would be content to come home to Hyannis and stay there the rest of my life to read, play bridge, talk to people around there and take a trip to Alliance now and then with a vacation once in a great while. Just to be back with real people."

Back in June Ty had written about the "seventh month of this thing." He contemplated what might happen to him. "The worst part is wondering if you are going to be caught and killed, or if not killed treated such that you lose your health and wish you were dead, or die because of it." He read a magazine story about an escaped prisoner who hid for four years. Ty figured he would need to hide for six months or more. "I don't think you believe I am dead, but that you are mostly worried about my health. I also expect that you think I am in a concentration camp. I am about as confined as if I were in one, but my outlook is much different and I am a thousand percent better off than if I were in one," he wrote the next week. By the start of the eighth month, he thought about how his life had changed so much since December 1941. "Then from April 9th to May 1st time didn't count. Part of that time, I didn't care what happened next, and time here the last three and a half months hasn't been too bad. Well, four more months, and I am hoping for another change." In December 1942 he got a change, but not the one he had hoped for.

IMPERIAL JAPANESE ARMY

1. I am interned at—Philippine Military Prison Camp No. __1__
2. My health is—excellent; good; fair; poor.
3. Message (50 words limit)

Everything is fine. Was very happy to receive your package. It came through in good shape. You couldn't have made a better selection in the things you sent. Am hoping for the time we will again be together. Give best regards to the Kokjers, Cobbs and our friends. Love.

Signature
Madison C. Kokjer

A half-dozen POW postcards made their way to Nebraska from the Cabanatuan POW camp, also known as Camp No. 1. Ty thanked his parents for a Red Cross box: You couldn't have made a better selection." He rated his health as "good," which meant the card made it past Japanese censors. Had he said fair or poor, it would not have been sent. Photo by Joseph W. Brown

Ty wrote a letter to his parents the day before he and more than 1,600 prisoners of war were loaded onto a hellship to be sent to Japan. A censor blacked out what Ty wrote as his likely destination—"to Japan." The letter contains the last words Hans and Shy heard from their son.

American airplanes bombed ships in a convoy December 14, 1944, the morning after they left Manila. The Oryoku Maru, which carried 1,619 prisoners of war, was hit twice. Ty and the other survivors were forced to swim ashore. Bombed again a few days later, the ship sank in Subic Bay with bodies on board. Naval History and Heritage Command, Fri., Dec. 15, 1944, National Archives and Records Administration

Chapter Eleven

Rumors Abound as the Safe Haven Disappears

"Ty, This notice dimmed the light of our hopes."

—Shy Kokjer, upon learning Ty was listed as missing in action, October 4, 1942

It was so peaceful where Ty was, it was sometimes hard for him to imagine there was a war going on nearby. Hans and Shy were addicted to news on the radio and in the newspapers, while most of what Ty heard about the war was rumors delivered by members of Filipino families who saw what was printed in the newspapers controlled by Japan, and what neighbors and friends in Guagua told them. Ty heard that 20 million Chinese troops would arrive to fight Japan by August 12, surely one of the more far-fetched war rumors. In November two Americans who had been hiding with Filipino families came through Pampanga on their way to join a guerilla group, bringing with them false rumors that U.S. forces had landed on Mindanao or Mindoro.

Ty heard true stories mixed with rumors about Japan's only other invasion of U.S. soil, two far-flung Aleutian Islands in Alaska territory that are closer to Tokyo than to Juneau. Ty thought the Japanese invaders had been repelled in June 1942, but that was when Japan landed troops, which he learned a few weeks later. The attack, meant to be a diversion from the Battle of Midway, did not result in the loss of all of Japan's ships, as Ty had heard. It took more than a year and scarce resources to retake Attu and Kiska islands. Similarly, the rumors he heard about the United States taking back Guam and Wake were vastly premature. U.S. forces did not retake Guam until August 10, 1944. Wake was bypassed; its return to the United States came when the war ended. Ty also heard the number of Japanese soldiers in Guagua, the nearby market town, was down to five hundred from three thousand.[165]

Filipinos sometimes asked Ty what happened to the U.S. planes that never arrived, why the United States surrendered instead of winning, and why no aid had been delivered. For these questions, Ty had no answers. He wrote in late June that he knew there were people worse off than he was and that many families would suffer when their loved ones didn't return. He also had some sympathy for the economic and other troubles of the enemy, in which he appeared to allude to more than Japan and Germany. "I know that our so-called enemies have their own troubles with lower standards of living, trade tariffs ... that we set up so our merchants can compete and keep our standard of living up. We can really only see our side of it even though we try to analyze the situation."

Their peaceful existence was occasionally interrupted. Ty was awakened late at night July 25 by a ten-to-fifteen-minute exchange of rifle and machine-gun fire, followed by another burst of gunfire in the early afternoon. Japanese and guerilla groups were fighting near Macabebe about three miles away. Several members of Amando's family fled from the village to the farm. "I am not exactly worried that we will be affected here, but it gives you kind of a feeling of being on edge," Ty wrote later that day. "I think this guerilla stuff is kind of silly as it causes such a lot of disorder when no good can come of it." There was some fighting the next night, and two days later, he heard that the situation in Guagua was grave, with guerillas trying to kill the pro-Japanese mayor.[166]

After the nearby fighting had calmed down, two American members of the 33rd Quartermaster outfit came for a visit. Their unit had arrived in the Philippines a few weeks before Ty did, and the men were living with a family up the river. Ty often ran out of things to talk to Bill and Tom about, so visiting with other U.S. soldiers was a welcome break in his routine. They also brought news—that there was fighting in Borneo. "I hope it is true," Ty wrote. It was, but it wasn't until 1945 that the Allies retook it.

In a Japanese newspaper dated July 24, he learned he would supposedly be free to go home by December 1, 1942, because the United States had suffered crushing defeats in the Pacific and could not defend itself. Australia, the newspaper argued, could save itself only by separating from Great Britain and joining "Greater East Asia." The paper also wrote about the American citizens of Japanese descent who were interned in the United States and said 2,300 American civilians were interned in the Philippines. The number interned civilians would grow to 3,700 by the end of the war.[167]

Over time, Ty's security and feelings of safety began to decline. On August 1, some of his host family members left for Guagua, "as they got scared out by the rumors." Things weren't much better for Ty. "The situation is rather grave today. The Japs are supposed to be making a house-to-house search for guns and guerillas. If it's true or not, I don't know. However we are sitting tight to hear something authentic. I sure wish I was in the sandhills of Nebraska at present." He could see smoke from houses that were burned about five miles southeast of where he was. The neighbors were praying, and he said, "I at times offer a prayer. It makes me feel better." In the end, he wrote, "Oh, a hundred years from now, I will never know the difference. I am trusting that my good fortune will remain with me and I will once again be with you." In a P.S. he said he had played twenty-five games of solitaire and he could still see burning houses. "This sitting here really gets on one's nerves. You are so darn helpless." Two day later, he had calmed down. "I don't think I will be upset by rumors anymore. ... The rice is growing and so when it isn't a field of water, it is a nice fresh green field." He found other moments of peace as well, writing in early September, "I feel almost like Robinson Crusoe sitting here in a pair of shorts, bare feet, dog at my side and cat on my lap." He added it was a year ago he had taken a cross-country flight to Dallas and was trying to get time off to visit home.

Other guerillas, whom Oming identified as radicals, came through the area, making Ty jittery. He hoped he wouldn't have to leave the farm and lose the notebooks made up of letters to his parents, which would compound his regrets about losing the previous diary in April. "But I guess it is all in war. I am certainly glad you are where you don't have to see all of this. I don't know that I am really afraid to die. It is just that life holds so much more that I would like to live to take part in it. I want to live more for the reason of being with you than anything else. I know how happy you will be when you again hear from me and know that I am all right. I will surely get through this O.K., and it will be really wonderful not to have to worry again."

Ty wrote about having a cold in August and wished he could be home to have his mother take care of him. He wrote about Pete, who had beriberi, for the first time in months. Their former roommate couldn't walk, even with help. Ty and the others had been sending fruit and vegetables to him. (Beriberi is a B_{12} vitamin deficiency, so beans, pork, and fish would have been more helpful.) "He wouldn't have been so horrifying if he had been dead. Eyes and mouth wide open. He is the skinniest person I have ever seen. Joints swelled. He had gotten himself all dirtied up with his fits this morning, and his clothes were off. All you could do was be thankful that he was unconscious and hope that he would pass away soon, as I know he will never get better and is bound to die. ... I wonder how those in the concentration camps are as I know they aren't getting much food and aren't getting the care Peter had." Pete, who served in the 31st Infantry, died the next morning.

A typhoon blew through from September 28 to October 2. "Half of Manila bay and part of the China Sea must be covering lower Pampanga today. The wind blew the water up the rivers, over the dikes, and the rice fields are now on a level with the river which is about five or six inches over the dike. ... I have been swimming twice today. If you walk, you wade in water over your knees, so it is easier to swim, even when going against the current."

After the typhoon ended, there was a shortage of food, which led to a discussion about going to Manila and either obtaining fake papers that would allow Ty, Tom, and Bill to be free, which seemed unlikely, or turning themselves in to become POWs. Ty rejected both ideas, saying he didn't want to make a move until he knew American forces had arrived. Later, Totung told him the regional governor said they would be put in jail if they turned themselves in. The local police suggested they go into the mountains, where many guerillas chose to hide. The guerilla encampments were away from main population areas, easily moved, and reachable only on foot. One mountain range runs down the middle of Bataan. Two others are just inland from northern Luzon Island's east and west coasts.

As Ty's situation was becoming less stable, Hans and Shy received a letter from the Adjutant General's office in Washington, dated October 4, 1942, saying what they must have suspected—Ty was officially missing in action. "Ty, This notice dimmed the light of our hopes," Shy wrote in her scrapbook. "I have thought of you always." They decided to believe he was a prisoner of war and hoped Japan would soon release a list of POWs. She pasted a news clipping following that news in the scrapbook without the newspaper name, although it was likely in Phelps County where Funk and Holdrege are. The story called Ty "a boy we are much interested in." It continued philosophically about war and soldiers' fates. "There is always a gleam of hope in that word 'missing,' as sometimes the missing turns up very much alive. ... We must steel ourselves to just such news, and worse, in this war. ... There will be many killed and missing when the most stupendous armies the world ever saw, meet on the battle field, as meet they must if the war is ever to be ended. ...Ty is the grandson of our townsman, U.S. Cobb, and a fine young man. ... There was a little air activity towards the end of the Bataan battle, and we are hoping he didn't crash in the Pacific."

October was a nervous time for Ty and the others. There was another firefight near Guagua and several other towns. With little hard information, Ty assumed it was a guerilla group that had demanded the withdrawal of Japanese troops by a certain date. "So some knot-head commander who is looking for glory has decided to start working on the Japs with the idea that aid will come very soon. If they have 150 guerrillas together they are doing good, I think." The next morning, October 14, "some joker" came across the river warning that Japanese soldiers were on their way. Ty grabbed his orders and his parents' letters, put on a shirt and pants, and swam across the river as if he were being chased. Doming gave him a gun, but it didn't work because it had been hidden in the river. When the scare was over, Ty returned to "a very fine" late dinner that included pork chops from a recently butchered pig. They could see smoke from burning houses across the river. He blamed the Japanese for the fires and the previous day's guerilla attacks for the provocation. They were on alert again the next day and heard machine-gun fire. He noted that any day he didn't write a letter probably meant he had been hiding in the fish pond.

On October 18, the "good friend" brought "our means of existence" and retracted his advice to turn themselves in to the Japanese. He said enemy soldiers were looking for guerillas in the mountains. "This is a darn good place here if the guerrillas will be peaceful, but if they are going to continue to raise Ned, I won't enjoy it here so much." A month later the good friend was "getting pretty scared, and I can't say that I blame him. I know I would be in a cold sweat if I were in his place." The previous night, some guerillas stayed near them. The group had fifteen pistols, two machine guns, at least ten thousand rounds of ammunition, three cooks, and six bancas. He thought they were the best-outfitted guerilla group he had seen, but not good enough. He hadn't written much the day before because the guerillas were looking over his shoulder and asking questions—about his family, if he was a millionaire, if he would take them to the United States after the war, and if he wanted them to find him a wife. Ty must have asked them as many questions as they pestered him with to learn about the group's strengths. On October 24, Ty again spent a hot morning hiding in the fish pond. "Another day I wish had never been."

Machine-gun fire one night made Ty jittery until he found out it was fish pond owners scaring away poachers. Rumors flew that Japanese soldiers would guard fish ponds, and "there might be raids in the next few days as the Japs were mad because the guerillas hadn't surrendered." Bill's wife, Nina, said she heard that someone was going to turn them in. Ty concluded it was time to stop mentioning names in his letters. The Japanese-controlled local newspapers were reporting victories for Japan, which Ty hoped were short-term wins. Getting home again seemed a long way off. November 8 marked the start of the twelfth month of the war. Writing on Armistice Day 1942, Ty said he hoped, without much optimism, the war would end that day in its honor.

That month, he heard the first war news he believed, from a radio broadcast that Totung translated. After losing territory in New Guinea, Australia began to get it back as of November 12; that battle would last until 1945. The Allies had a victory in Casablanca in November. Spanish dictator Francisco Franco made a public declaration of neutrality, although he secretly pledged support to the Axis powers. On the downside, Russia was bogged down at home. MacArthur moved his headquarters to Port Moresby in New Guinea in November, the start of a long and costly battle and the march north toward the Philippines. American and Australian forces were engaged in heavy fighting there in December. The fight continued through the end of the war, with heavy casualties to soldiers and the local population. A Japanese ship hit a U.S. mine and sank near Tokyo on December 20, the first successful attack in Japan's home waters.[168]

Ty heard that Roosevelt said U.S. production capacity would win the war and that, by 1943, the United States would have 9.7 million men in the armed forces. Those figures were close. In two speeches, one on September 7 and one on Columbus Day, the president projected the country's might. War production plants were "getting ahead of the enemies," Roosevelt said. Almost 2 million men were in the military, up from 333,500 in 1939. By 1945, the armed forces would include 12.2 million men and women.[169]

In mid-November, Ty wrote about an unusual evening, at least for him. Totung brought a bottle of gin over, "and Bill, Tom and I had to try it." Tom gave up after half a glass and went to bed. "I drank about two glasses, got sick and threw it up." Bill drank about twice as much and was sick the next morning. Ty partly blamed the lack of ice, and the "Filipino pop we were using for a mix was no good."

In his last diary entry for November, on the 18th, Ty was nervous. "Maybe I am a little more scared than Bill or Tom, and maybe I am lazier and am content to spend my time in the house. ... I have been here nearly seven months now and the time has gone fast and nothing bad has happened. And things should be no worse for the next six months. Maybe it will end sooner (I hope)." As he ran out of paper, he said he would not bury the notebook he was writing in. (He did bury the note-book ending November 18.) "I can't help but think things will come out all right in the end, but there is always the possibility of the unexpected to happen. Why things have to be as they are, I don't know. But I guess this life was meant to be full of mysteries. I guess I have benefited a great deal from the time I have spent here and will spend here, but it could have been much more pleasant in other places. ... I want you to know that I have been and will be thinking of you and the better things in life."

When the diary resumed December 4, Ty wrote about seeing fourteen Japanese dive-bombers overhead. It was not unusual to see Japanese planes, but this time they were bombing sus-pected guerilla camps nearby, so he and the other Americans hid in the fish pond until dark. He ate dinner at "my friend's house" and stayed until midnight watching them cook for everyone who would participate in rice planting the next day. He "really enjoyed the evening." The next day, "planes were at it again" and guerilla forces retreated in their direction. Ty didn't go to the fish pond, but, "Everyone was scared spitless." He was at the house of someone he didn't name and met some women who said guerillas were in Ty's bamboo hut. "And don't think but what I was in a cold sweat all the rest of the day." Two men, Fred and Bert, and a third, a member of the 27th Bomb Group from Mount Vernon, Texas, whose last name was Lovelace, visited the house. The three of them warned him about guerillas wanting to take Americans with them, by force if necessary.

When Ty thought it was safe to return, he discovered the guerillas were still there. "They got me cornered and said, 'You don't have to go with us but if you don't want to go, you have to talk to the Major,' who I found out to be Sergeant Guito (or Guinto). Well I didn't go and I didn't talk to Guito either. That man has caused and tried to cause us more trouble than everyone else combined." Ty slept next door that night. It's not clear why Ty was angry with Guito or if he was a member of the local police force, but it's a probability. Guito brought Ty and his roommates food at least once, but he more often brought rumors.

The next day brought more rumors about Japanese soldiers searching for guerillas. Ty hid in the fish pond in part "from fear the guerillas would be back to kidnap us. Which we later found they planned to do, but the Japs stopped them." While he was gone, the only razor the three Americans had, shoes, cigars, a pipe, tobacco, and medicine disappeared, presumably taken by the guerillas. "The Japs call them bandits, and that is about right. Some of them are there for a noble cause, but the leaders and the majority of them are socialists, communists, or have served time in prison. I don't want any part of them." Some of the "bandit" groups arose when most members of local constabularies were drafted into the Filipino army, leaving small communities without protection.[170]

That week in her scrapbook, Shy pasted a story from a Lincoln newspaper, based on Japanese reports from the Philippines: "Strong, completely organized and well-equipped United States and Philippines army forces are still holding out in the central Philippines, eight months after the fall of Bataan and seven months after the surrender of the heroic garrison at Corregidor, Japan disclosed Tuesday in a series of broadcasts." It wasn't clear to Americans why Japan was all of a sudden broadcasting news about the Philippines. But the news story said the guerillas must be under control of high-ranking officers because of "its evidently splendid organization." That may have been a reference to Colonel Claude A. Thorp, designated by MacArthur to organize a guerilla operation and funnel information to Australia. Thorp, who led a large group, was captured in October 1942, along with some of his men. They were executed in 1943. Ty knew about Thorp in October, but was skeptical of his reach. In November 1942 Ty heard rumors about Thorp's capture.[171]

The ten days from December 7 to December 17 were a whirl of rumors, terror, and acts of kindness toward the Americans. On the morning of December 7, "Things started in full force to the east of us and moved northwest. Planes, bombs, machine guns, mortars." More neighbors left for Guagua as people whose homes were being bombed retreated in the direction of Ty's hut. The next day started with more bombings, then things went quiet. "The Jap infantry and their constabulary moved in, I guess." This couldn't have been good news, although Ty didn't express any regrets about remaining where he was instead of heading into the mountains or looking for his guerilla contacts.

Bill and his wife left. As a Filipina, Nina likely had friends or relatives she could turn to. They didn't go far, however, because Ty and Tom spent some time at their home, and Bill brought them a "coconut sweet" one day. With the family that had cared for them in Guagua, Tom and Ty had little money and an irregular supply of food from neighbors. They continued to hide from both Japanese and guerilla forces. "That night everyone took off for Guagua and we slept in our hiding places around the fish pond." They stayed there, with occasional forays for food. Sometimes they hid in a neighbor's house.

The morning of December 8, he heard bombs nearby and could see houses burning to the south. "Well, we are now in the second year of the war. I hope this turns out to be an 18-month war like the last one," he wrote December 9. "Just saw a flight of planes go toward Clark [air field]. Couldn't tell how many. They haven't been flying in this area this morning, which makes me very happy. I hope the people around here come back such that Japs won't come in and burn their houses." He took short walks most days, but usually hid in the weeds near the fish pond. The next day, while running from a report of Japanese soldiers nearby, Ty tripped and fell into the pond. As a result, "I am stiff, sore, have a headache, nose is running, and gas on the stomach, but still in good spirits."

Ty made several references to "my Filipino friend" over the next few days. "She gave me a small picture of herself and also one of her handkerchiefs." He hadn't professed admiration for a local woman for some time, and it's not clear who the friend was. Neither of the gifts survived the war. He also wrote about "my helper." To mask her identity, he called her Betty. She and her brother were visiting December 11. She cooked dinner for him; Tom and Bill were gone. Betty was staying next door with a cousin, her brother, and two other brothers. "About dusk, we saw many fires towards where Betty lives, and her youngest brother went to investigate. He came back after dark and was crying that they burned their home and had taken their parents to Macalbebe [Macabebe] as the police central was looking for a brother who was a guerilla. Betty really went to pieces and I slept in the fish pond again."

He was advised after breakfast on December 12 to go hide where Tom and Bill were, which he did. He played bridge with Tom, who won two out of three rubbers. Bill feared their presence could endanger him and Nina, and he told them to go back to the bamboo hut. "I hate to cook myself, etc., but it was almost a relief to be away from Bill and his wife."

He and Tom had seven eggs, meat for two meals, rice for two weeks, lard for a week, sugar for a couple of weeks, and two pesos. There was no one left nearby to buy anything from, so the lack of money didn't matter. He had acquired a .45

automatic and a dozen bullets. If he had a head start, he said he wouldn't use it. "However, I hope I am killed fighting, rather than captured and then be tortured a while and end up being beheaded." He said he was "perplexed" and wasn't in the mood for the holidays or his birthday. "I wish all this were past so I could be back to normal life. I have had about all the adventure I want." Local fishermen gave them several fish, but a dog or cat absconded with half of them overnight. Neighbors brought supplies "and there went the rest of our money." They ate fried pork and french-fried sweet potatoes.

Two events that would shape history occurred in strict secrecy that fall: the first U.S. jet plane, the Bell XP-59, flew across the Mojave Desert on October 1, although jet planes did not fly in combat in World War II; at the University of Chicago, the first nuclear chain reaction took place December 2, foreshadowing the end of the war three bloody years later. On December 17, British foreign secretary Sir Anthony Eden informed Parliament that Hitler was carrying out his promise to exterminate the continent's Jewish population. Ty missed that bit of news, but he did hear a rumor that U.S. forces had landed on Sicily. In reality, that didn't happen until July 1943.[172]

One day earlier, on December 16, Ty wrote that he and Tom had food, and things were calm enough that they played three rubbers of bridge, and Ty read about India in *National Geographic*. But that was the end. His last diary entry, and the last uncensored words his parents later read from their son, were written as they were captured December 17, 1942, six days before Ty's 23rd birthday.

"Well, it happened this morning that what we have been fearing for some time." Tom cooked breakfast, and then a policeman from Macabebe showed up to take them to his boss. After dinner, they started for Macabebe in a banca to be turned over to the Japanese. The police chief said he would keep Ty's last diary notebook and other things until after the war and send them to his parents. The bulk of the diary remained buried, but the chief survived and kept his word.

This is the final entry in Ty's diary:

I don't know quite what to expect, but I am more or less expecting the worst, as they have every right to shoot an escaped prisoner of war. I just hope they don't torture me. I guess this may be the last you will ever hear of me. You know how much I love you and know that you will miss me, but I believe that if I am to die this way, it may be for the best. I still have hopes of returning to you, but they are very slim.

These people who have captured me hated to, I know, but it is their duty and just one of those things of war. Now, at least, we will cause no more trouble to the civilians. I hope the Japanese do not force us to tell who has helped us as they were innocent people. The day is nice with clouds in the east. But they will clear in time as will the war—clouds and things will again be normal. I would like to return to the States, find Mary, and live the life I was supposed to live, but I am glad I have no dependents. You have given me all a boy could want and I only wish I could live to repay you. But it looks as though my fortune is not to be such. Look on this Chief of Police as a friend as he is not doing what he likes to do, but as he had to do.

With all my love and thanks to you.

Your loving son,

Ty

P.S. The Chief says he sends his best wishes, and that after the war, he will write if he lives through it. He is a Filipino soldier who was captured, and then forced to work under the Japanese.

Ty and Tom spent a few days in the regional Pampanga jail before they were transferred to Bilibid Prison in Manila in January 1943. Tom remained in Bilibid after Ty was sent to the Cabanatuan POW camp in early February 1943. Tom was liberated in early 1945, and he wrote to Hans and Shy in May of that year, recalling what had happened after their recapture. By then, Ty's family was fairly certain that he had been put on a ship and was likely in Japan.

"I'm glad Ty is safe and I'll write to him," Tom wrote from a hospital in Michigan, near his home in Detroit. "Ty was in good health at all times except for a case of jaundice in June 42." Jaundice can be a symptom of malaria. Later, a friend reported that Ty, like many POWs, had malaria. Tom's letter filled in some other gaps. After being recaptured in December 1942, they were taken to the "Provincial Prison of Pampanga at San Fernando. Here we had rice and fish. ... We went to Bilibid on January 9, 43 and received 2 Red Cross boxes."

Ty wrote more than once that he would rather die than be captured and tortured. Based on Tom's letter, it seems unlikely they were tortured at Bilibid.

Another clue about what Ty endured after his recapture can be found in the story of his flight school classmate, Robert G. Bjoring. He and Ty arrived in Manila the same day, and they were together the night of the surrender in April 1942. Bjoring and three other men decided they would try to evade capture, but Ty refused to go with them.

Over several months, Bjoring's group headed east and south on Luzon Island, paying guides and taking temporary refuge with Filipino families. Their plan was to catch a supply boat that would take them from the south end of Luzon, many miles south of Manila, to Mindanao Island. From there, they would find a way to get to Australia, as a few Americans did. Japanese soldiers discovered them near the boat dock, and Bjoring was taken to Bilibid Prison and questioned. He said he was beaten a few times, but thought he got off easy. He had been out of touch with other Americans for so long, he figured that Japanese officials decided he didn't know anything of value. Bjoring was sent to a POW camp in Japan in late 1942 and survived the war. If Ty had joined them, it's possible the group dynamics would have changed their destiny. But it's also possible Ty would have been sent to Japan in 1942, when ships were rarely bombed.

Thousands of POWs were moved throughout the Pacific theater, mostly to provide labor where Japan needed it, including on the infamous Thailand-Burma railroad. POWs were enslaved at Japanese farms, factories, mines, and shipping docks to replace men who were in the armed forces. By the end of the war, Japan had captured 132,134 POWs around the Pacific, including those in the Philippines. More than a quarter of them died. The death rate for Americans was 33 percent. By comparison, Germany and Italy held 235,473 prisoners with 9,348 deaths, or 4 percent.[173]

During Ty's early days back in captivity, the United States and Britain discussed the timing of a cross-channel invasion of France at the Casablanca conference. Roosevelt announced a policy requiring unconditional surrender by Japan and Germany, likely prolonging the war.

From the start of the war until May 1943, Ty's parents knew little of his whereabouts or condition, but there were occasional snippets. It was October 4, 1942, when Hans and Shy learned Ty was officially missing in action. They never suspected he was well and in hiding. They forced themselves to believe he was a prisoner of war. Those hopes must have dimmed when a January 1943 newspaper story reported that few members of the 27th Bombardment Group were still alive. Not until May 25, 1943, five months after Ty was recaptured, did they learn he was a prisoner of war. "This is an eventful day dear Ty, a message came from Washington. You are alive, that is the joy that has come today. Thank God we still have you. Without you our world would be so dark," Shy wrote in her scrapbook, adding that the news was in the *Omaha World Herald* the next day.

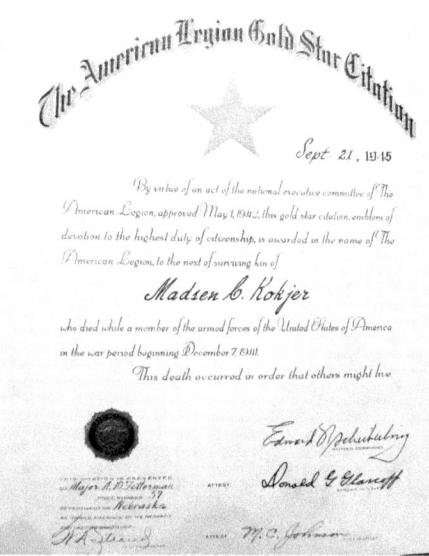

Hans and Shy received this American
Legion Gold Star Certificate dated
September 21, 1945. It reads, in part,
"This death occurred in order that
others might live."

Hans and Shy visited Japan in 1958 with a group of Americans. They met with a professor and toured Moji, where the 450 survivors of the Oryoku Maru docked at the end of January 1945. Ty died there February 12, 1945. Family photo

Hans, second from left, and Shy, fourth from left, visited Ty's grave at Manila American Cemetery, probably in 1958. With them, left, Dr. Jose Jose; third from left, Mely Jose, wife of Domingo "Doming." Far right is Enrique "Iking" Jose, one of the brothers, and the tallest member of the family. Family photo

Chapter Twelve

Life and Death on a Hellship

"I have big plans for after the war."

—Ty Kokjer, December 13, 1944

Ty had been well fed for eight months, followed by poor rations for six or seven weeks, when he arrived at Cabanatuan prisoner-of-war camp in February 1943. As a result, he was likely in better health than his fellow POWs, who had been prisoners for nearly a year. He also arrived at the camp when the death rate had declined. Several members of the 27th Bomb Group and Ty's pilot-training classmates were there, and as usual, Ty would have searched for familiar faces and made new friends. Postwar letters from those who knew Ty provide a few details about the twenty months he spent at the camp, in Luzon Island's central plains, about seventy miles north of Manila. Like Pampanga, it is a tropical, rice-growing region.

Those who completed the Death March in the spring of 1942 went first to Camp O'Donnell, which some prisoners called "Camp O'Death." It was an unfinished—and woefully unprepared—training facility intended for Filipino troops. The best estimate for the number of U.S. and Filipino POWs who reached O'Donnell is fifty-four thousand. Like Ty, some escaped, and thousands died on the march, more than had died in battle. Two days after the first of the estimated nine thousand American POWs arrived at O'Donnell, a Japanese-controlled Manila newspaper reported that camp officials were going out of their way to take care of the captives, adding it was the fault of the United States that so many of its soldiers were weakened by disease and the lack of food. "The Death March had been hell, but O'Donnell was a new kind of torment," Manny Lawton, a captain in the 31st Infantry, wrote in his memoir. Three slow water spigots served the entire camp, and prisoners fought for water or waited in line for up to twenty hours to fill their canteens.[174]

In July the prisoners were transferred to the somewhat better Cabanatuan, about sixty miles to the east. "At first it was nothing but O'Donnell continued," historian Gavan Daws wrote. At the rate soldiers were dying in the summer of 1942, some figured they would all be dead in ten weeks. The American POWs were housed in thatched-roof barracks designed to hold 40 men. But up to 120 were crammed into each, where they slept shoulder-to-shoulder on wood beds. They became infested with roaches, fleas, lice, and bedbugs. In the chow line, men groomed each other like monkeys, picking lice off each other's backs. The first day that no one died at Cabanatuan was January 18, 1943, about three weeks before Ty arrived. By the summer of 1943, the death rate was down, likely because POWs had built sewer and septic systems. Soldiers set up rain barrels, improved latrines and drainage around the barracks, and used lime to control flies and maggots, bringing down the dysentery rate. Baths were rare, and prisoners showered when it rained by taking off their clothes and standing under the eaves of buildings. Few had soap or toothpaste. Some men chewed the ends of sticks to make toothbrushes.[175]

Cabanatuan and a prison farm on Mindanao Island were the two major POW camps in the Philippines. They were, essentially, holding camps. Japan had other uses for POWs, including rebuilding roads, bridges, ports, and runways around the Philippines destroyed by Japan's own bombs. Beginning in late 1942, Japan moved prisoners around the Pacific to work as laborers. Prisoners were moved from Luzon to Mindanao or Palawan islands. The first so-called hellship to leave the Philippines took 180 POWs from Manila to Formosa on August 12, 1942. From the time hellships began transporting POWs, author Gregory F. Michno counted 156 voyages on 134 ships. Except for traveling on the ocean instead of railroad tracks, the hellships resembled the boxcars that Germany used to transport Jews to death camps.[176]

Ty and the healthy prisoners, a relative term, would eventually be sent to Bilibid Prison in Manila to await their voyage to Japan or elsewhere in the Pacific. In the meantime, they did the best they could at whatever camp jobs they were assigned. A report in the *Grant County Tribune* in September 1945 said Ty worked in camp administration, but there is no other mention of what work he did. Most POWs farmed vegetables using hand tools. Some were cooks, some gathered firewood, and others did camp and latrine maintenance. Some worked on the docks in Manila, rebuilt roads and airfields, or took disabled American tanks and trucks apart for scrap metal. One POW, who owned a jewelry store at home, repaired watches, clocks, and radios for Americans and Japanese guards. He salvaged enough parts to build a secret radio so POWs could listen to news reports. In Manchuria, Wainwright raised vegetables and herded goats.

The Army notified Hans and Shy in July 1943 they could send a package to Ty through the International Red Cross and provided a list of things they could include. The box could weigh no more than eleven pounds. Vitamins and concentrated foods were at the top of the list, with a reminder to avoid anything that could not survive high temperatures. Among the recommended items were medicated soap, towels, tooth powder, toothbrushes, safety razors, razor blades, nonbreakable mirrors, styptic pencils (used to stop the bleeding of small cuts, often from shaving), playing cards, checkers and other board games, footballs, socks, belts, shirts, underwear, handkerchiefs, chewing gum, safety pins, combs, fountain pens, house slippers, low-priced watches, pencils, candy, processed American Swiss cheese, dried soups, powdered milk or malted milk, and sugar. Hans and Shy checked many of the preferred items on their list, although it's not clear if they were able to include all of them.[177]

Unbeknownst to Ty's parents, Japanese guards routinely delayed the delivery of Red Cross boxes and looted them, taking food and medicine, then lowered the prisoners' already minimal food portions to compensate for the canned meats, candy, and other food. American prisoners in Japan received fewer Red Cross boxes than POWs in Germany. It could have been lack of organization, poor transportation, or vindictiveness. Soldiers sometimes got intact boxes with quinine, Spam, cigarettes, canned milk, dried fruit, and chocolate.

Prisoners were occasionally allowed to write postcards to relatives, and six from Ty eventually reached the United States. Prisoners indicated their general health by ticking one of four choices on a list. If a prisoner said his condition was fair or poor, wrote he was sick, or wrote about true conditions for POWs, the card would not leave the camp. Prisoners feared if they checked excellent, they would be sent to Japan. But if they said their health was good, the card had a better chance of being sent. POWs could add twenty-five words (sometimes fifty), which Ty used to send his love to his parents, other relatives, and friends. American censors stamped numbers on the cards to show they had been approved for delivery.[178]

In August 1944, Hans and Shy received a postcard dated May 6, 1944. Ty said his condition was good. "Everything fine. Best hopes for a happy birthday Dad and Gramps. Always thinking about you Mother. I have mail from you. Time passes quickly and we will be together again. Enjoyed your box very much. Best regards to our friends and relatives. As ever, your loving son." Instead of "Ty," he was required to sign his full name on the card. Shy's birthday was in July, and Hans shared a November birthday with his father-in-law, so the timing of the wish is odd. In another card, this one undated, Ty's health was good and he thanked them for the Red Cross boxes. A card dated July 15, 1944, had a shorter message. "Getting along alright. Hoping for a time we will all be together. Enjoy letters from home." It arrived in Hyannis January 24, 1945, and was forwarded to Funk, where Hans and Shy stayed with her father for a while after giving up the hotel.

The postmarks on some of Ty's cards are unreadable, and more than one was delivered after the war was over. After arriving in Hyannis, the cards were forwarded to Hans and Shy's apartment in Lincoln. One card that was sent to Hyannis, then to Funk, carried a September 14, 1945, postmark from Hyannis, but there is no indication when Ty wrote it. He said his health was good. "Everything is fine. Was very happy to receive your package. It came through in good shape. You couldn't have made a better selection in the things you sent. Am hoping for the time we will again be together. Give best regards to the Kokjers, Cobbs and our friends."

Hans and Shy were allowed to send short messages to Ty on postcards or air-mail stationery that folded to become an envelope, both with a word-count limit. He received some, but others were returned, some after the war ended. A postcard from Shy dated January 26, 1945, was sent to Ty as "Interned in Japan." She wrote: "Your cards make us so happy. Irvy sent elk to Gramp, and Thomas sent wild duck." The date May 15, 1945, was handwritten on the front, likely indicating when it was returned.[179]

In an undated form letter, Hans and Shy wrote, "Each day nearer reunion. Happy health good. Ours excellent." One from Hans sent June 26, 1945, came back in early January 1946. Hans had written that Ty's cousin Bobby Cobb was back home with his wife and baby boy, adding, "Mother, Gramp I fine." As they often did, Hans and Shy shared whatever news they received about Ty with friends and family, and Shy wrote in the scrapbook that the whole town rejoiced with them. In the camps, "A man with just one letter would read it over and over," Daws wrote. Many letters sent to prisoners disappeared. Wainwright, who was imprisoned in Manchuria, received six letters from his wife, who had sent three hundred. He was allowed to write three letters to her.

A few camp commanders treated POWs relatively well. "Most, however, ranged from average to horrible; some couldn't care less, others were drunks, some were sadists, and a few were homicidal ... and often a prisoner's fate was decided with no more rhyme or reason than in the proverbial luck of the draw," Michno wrote.[180]

POWs stood in the hot sun when they arrived, waiting for the commander's speech and dozens of rules. David L. Hardee reported a translation of the speech when he arrived at O'Donnell on April 25, 1942: "You are my enemies. I should like to kill every one of you, but you are the property of my emperor and my Bushido forbids it. I should like to carry you to my country and let you see the great damage your embargos have done to us. We are at war and the hatred you have caused will not be forgotten for a hundred years."

POWs were assigned to ten-man squads. If one member escaped, the other nine would be shot. Dyess and Army Captain Charlie Underwood recalled three officers who were caught trying to escape. The men were beaten, tied to poles, and left in the sun and rain for two days while they were beaten and whipped, then taken into the woods to dig their own graves before two were shot and the third beheaded.

Japan's war minister, General Hideki Tojo, issued an order soon after the surrender that POWs had to work if they wanted to eat. An estimated two thousand five hundred prisoners raised vegetables, including beans, carrots, peppers, corn, onions, okra, sweet potatoes, and spinach. Most of the crops went to Japanese soldiers, and some of the crops were traded for rice. The POWs occasionally got small amounts with their allotments of rice. Once in a while, they got an ounce of carabao, a bite of chicken, or a bit of an egg, but not enough to stave off nutritional illness or weight loss.[181]

Sometimes, protein came from the weevils and worms in their meals. John S. Coleman Jr., a captain in the 27th Materiel Squadron, wrote in his memoir that one day he counted eighty-two insects in his cup of rice, which he ate. Men hated the unsalted rice and craved meat, sugar, and cigarettes. Some traded meals or gambled future Red Cross boxes for cigarettes. Before arriving at Cabanatuan, POWs lost an estimated average of fifty-five pounds, and many were near death. Fewer died as a diphtheria outbreak waned and medicine and extra food arrived at the camp, some through intricate smuggling operations. In 1944, only two POWs died at Cabanatuan. POWs were never given shoes or clothing. When they could, prisoners removed them from dead men.

In their search for information, Hans, Shy, and relatives wrote to soldiers, newspaper columnists, radio broadcasters, and other strangers. Sometimes they got false assurances about how POWs were being treated. Staff Sergeant Donald J. McPhee wrote March 27, 1945, that neither he nor his friends knew Ty, but he commented on photographs taken at Cabanatuan a year after the surrender, when Ty was there. "With careful consideration of the possibilities of propaganda, the pictures give a reasonable view of the average daily camp life up there. The men were living in fairly normal surroundings, and there should be no real anxiety about their treatment." He was right about propaganda. When Red Cross workers and photographers came to inspect, camps were temporarily "dressed up" to appear to meet international standards. Poor reviews could lead to jail or death for the local inspectors.[182]

Hans and Shy got other information as rescued POWs began to return to their hometowns in 1945, including two Nebraskans. Captain Merle "Jim" Musselman, an Omaha doctor who met Ty at Cabanatuan, wrote to Hans and Shy in mid-March, about a week after returning home: "He had suffered from chronic malaria and malnutrition, but was in good health when I last saw him. ... I do know that he loved to sketch and he showed me many sketches of the hotel. He talked about you and Hyannis and looked forward so much to his return. ... He had many plans for the aftermath." First Lieutenant Jack Obbink wrote from a rehabilitation hospital, after a quick visit to his mother in Lincoln, which the local newspapers wrote about. He was among 1,275 POWs rescued from Bilibid Prison in Manila in early February. He thought Ty had been a guerilla. In a letter to Dr. Albert N. Sarwold, dated April 20, 1945, Hans wrote about what Musselman and Obbink had reported. Ty and Sarwold had been at Bilibid together before Ty was sent to Japan. "We are happy to know that Ty was always cheerful and that he did have such good friends during these trying times. We are sure that we can thank you fellows for saving Ty's life."[183]

Shortly after Ty was recaptured in late 1942, the United States and its allies began to win battles against Japan, which was losing conquered territory and stopped from further advances. Japan's defeat at Guadalcanal in February 1943 is an example. "It confirmed for the American public and its fighting troops that the Japanese army was not the invincible machine that they had been led to believe," Winston Groom wrote. By early 1944 the United States had air and naval superiority in the Pacific.[184]

Japanese commanders in conquered territories weren't developing resources fast enough to suit their superiors. Japan's industrial output had gone up one-fourth, as the army and navy fought over resources. In the United States, output was up by two-thirds. Emperor Hirohito summoned Tojo, who had become prime minister, to ask when the war would end. Tojo later quoted the emperor: "You keep repeating that the Imperial Army is invulnerable, yet whenever the enemy lands you lose the battle. You've never been able to repulse an enemy landing. Can't you do it *somewhere*? How is this war going to turn out?" Japan's system for supplying far-flung troops began to fail during the Solomon Islands campaign, and in 1943 the Allies sank supply ships sent out without military escorts. "The first full year of the Pacific war ended with some notes of optimism for the Allies. There had been no sweeping territorial gains, but the Japanese Army and Navy had been halted. ... The Japanese could be beaten," Michno wrote.[185]

MacArthur fulfilled his promise to return to the Philippines on October 20, 1944, when he famously waded ashore on Leyte Island, two islands and 350 miles south of Manila. The furious battle of Leyte Gulf, October 23 to 26, is considered the largest sea battle ever waged. The Allied victory meant "never again would the Imperial Navy play more than a minor role in the defense of the homeland," historian John Toland wrote. The battle also marked Japan's first use of kamikaze planes, which carried bombs and pilots whose final, fatal missions were to ram American ships.[186]

On September 21, 1944, the United States began bombing bases near Manila, the start of a brutal battle for the capital city that ended in early March 1945, after one hundred thousand civilians had been killed. The bombings brought hope to civilian internees at Santo Tomas University and to POWs at Bilibid Prison. Ty may have been at Bilibid by then, but the bombings were also visible from Cabanatuan. The remaining healthy POWs at Cabanatuan were soon moved to Bilibid. "There was alternating laughter and tears," Manny Lawton, a POW at Bilibid, wrote about the bombings. "It was the most beautiful sight we had witnessed during the entire war; the first actual proof that our side was winning." Hardee, also at Bilibid, wrote, "The explosions were the sweetest music we had ever heard." A secret radio meant he and other POWs knew about the Leyte naval victory.[187]

The joy of returning American planes died for some, however, when they realized Japan still intended to put them on ships. "Morale sank to the lowest point since the fall of Bataan. ... Freedom seemed to be within our grasp. Now those dreams vanished," Lawton wrote. He was sent to Japan in December 1944.

Lester Tenney, a private in a tank battalion, was among five hundred prisoners taken from Cabanatuan to be put on *Toro Maru*, the first hellship to leave Manila for Japan in September 1942. The thirty-two-day journey included a two-week stop for repairs to the decrepit ship. Many of the POWs had malaria or dysentery, but none of them died, despite meager rations. "We realized that we were going to be relieving some of the Japanese citizens from certain menial jobs, against the principles of the Geneva Convention, thus freeing them for military service to fight for the emperor," Tenney wrote. They worked in a coal mine at the POW camp Fukuoka 17, on the Japanese island of Kyushu, the southernmost of Japan's large islands. The island's port is about halfway between Nagasaki and Hiroshima. The first trip from Manila to that port, Moji, Japan, a city near the country's main coal-mining industry, took 17 days in November 1942. Other early ships made voyages of two to three weeks. As bad as Tenney's journey was, later journeys were far worse. The death rate climbed as the war continued. "Nineteen forty-four would be Hell Year," Michno wrote. Of the 126,064 POWs transported on hellships, an estimated 21,039 died. The lack of precise records means the actual totals will never be known.[188]

With MacArthur's imminent return to Luzon Island, Japan rounded up the remaining able-bodied POWs and, on December 13, 1944, put them on the *Oryoku Maru*. Ty was one of 1,619 men forced onto that last hellship to leave Manila. Most of the POWs were taken from Cabanatuan. The rest were from the Davao POW farm on Mindanao Island. Ty's family knew he was alive in December 1944 and had been put on a ship. His Uncle Emerson Kokjer wrote to Ty's fraternity six months later, on June 12, 1945, to ask if the fraternity knew something the family didn't after the magazine reported Ty was dead and sent a gold star to his parents. "As your letter has naturally affected his mother and father very much, will you please send your reply by air mail immediately," he wrote. The fraternity updated its news about Ty in the next edition and asked that the gold star be returned.[189]

Further proof that Ty was alive in the fall of 1944 came from his cousin Chan Cobb, who was in the Navy and stationed in San Francisco when he wasn't at sea. After Ty and the others bound for the *Oryoku Maru* were transferred to Bilibid, five hundred POWs too sick to be taken to Japan remained at Cabanatuan. They were rescued in late January 1945 by U.S. Army Rangers after American forces landed at Lingayen Gulf about 135 miles north-northwest of Manila and 90 miles northwest of the POW camp. The Army feared the POWS would be killed once Japan realized Luzon was lost. Most of the rescued POWs were sent to San Francisco, some by plane. On February 7, 1945, Chan wrote home that he had met one of the men at a military hospital "who is very sure" Ty was sent to Manila. "Ty is pretty [likely] to have been in good shape at that time or they would have left him in the camp. Hate to find this information as it is but it could have been a lot worse. These men seem to think our boys will get better treatment in Japan." More false assurances came in a letter from NBC radio broadcaster Larry Smith. Isabel Kokjer wrote to ask him for more details about one of his news reports. Smith replied February 21, 1945, that men on the sunken ships had all come from Davao, not Cabanatuan, and all had been identified. Only POWs in good health had been relocated to other camps in the Philippines or taken to Japan. None of his assurances were true.[190]

Ty's name appears in records kept by Japanese officials, West Point, the MacArthur Memorial and Museum in Norfolk, Virginia, and other archives that tracked the *Oryoku Maru*. Many soldiers who survived the Death March, prisoner-of-war camps, and slave labor camps in Japan and elsewhere in the Pacific have written memoirs or have been interviewed by authors and filmmakers. This makes it possible to surmise what happened to Ty during this interval, despite a lack of correspondence from him, since he was in the same place at the same time as those soldiers.[191]

Before Ty left Bilibid to march to the dock, he wrote a note to his parents and gave it to a soldier who was too sick to make the journey. Edward Blaine Claypool, of Jackson, Mississippi, wrote to Hans and Shy from a U.S. military hospital April 26, 1945: "[Ty] was feeling fine and was in good spirits before he left, and I hope he will come through all right." Claypool was among 800 prisoners liberated in February when American troops reached Manila. He enclosed the note from Ty, written on a torn piece of notebook paper. A word, or words, was blacked out in dark ink different from what Ty used, presumably by a military censor, and presumably it read "to Japan." It was his last message to his parent.[192]

"Just a note as I guess we make a move tomorrow, [blacked out] or maybe just local. Anyway, the fellow I am leaving this with may be free before I am. I sent a card to you today. And it said my health was fair, that is down in weight perhaps to 120. I feel O.K. Have had a cold but am in pretty good shape and am sure I will make it O.K. I have big plans for after the war. The past three years haven't been too bad. A little light on the chow the last few months. I may have a chance to write more later. Till then or when I can communicate again, so long as ever. Your loving son, Ty." Unlike others who wrote about being starved, Ty didn't want to alarm his parents; instead he wrote about being "light on chow." (He weighed 170 pounds when the war began.)

The book *Belly of the Beast*, by Judith L. Pearson, is a detailed account of the journey from Cabanatuan to Japan. She reviewed diaries and talked to more than a dozen survivors and relatives of others who died some years after the war. Other books include tales of survivors, and Gregory F. Michno catalogued the story of every hellship: "Prison camps were bad, but hellships were worse." Some POWs compared the hellships to slave ships, and Michno wrote, "It does not take a great stretch of the imagination to find many similarities."[193]

The journey of the *Oryoku Maru*, a modified passenger ship, began ominously. As the POWs marched out of Bilibid Prison on December 13, 1944, there was an air raid. It was a moment of terror, hope, and disappointment for the prisoners, knowing they might be killed right then, and they were leaving for Japan as U.S. forces were about to begin the invasion they had longed for. At the dock, they waited as a few hundred Japanese civilians, the families of occupation government officials, boarded the cabins on the upper decks. Also on board were several hundred others, including guards, the ship's crew, and Japanese sailors whose ships had been sunk. Many jeered at the Americans. Guards included the cruel interpreter Shusuke Wada, who was later charged with war crimes and sentenced to life in prison. MacArthur's Packard was part of the loot loaded onto the ship.

Calculating the length of hellship journeys is complicated. Some prisoners were loaded onto ships that sat at the Manila dock for a day or two. Other ships left the pier but anchored in Manila Bay before sailing. Some stopped in Takao, Formosa, before offloading POWs at the port of Moji near Japan's coal mines on Kyushu Island. At a mine in Yamaguchi province north of Fukuoka, POWs worked twelve-hour shifts for ten days before getting a day off. At the end of each day, they were covered with coal dust, which also filled their lungs. If a guard thought they weren't working hard enough, they risked a beating.[194]

The journeys of the first hellships that left Manila for Japan in 1942 and 1943 generally took two to three weeks. Some trips were longer as ships were damaged, and others needed repairs. Later, as the Allies had more planes, ships, and submarines in the Pacific, convoys zigzagged or hugged the coast of China to avoid attacks, slowing their progress.

The *Noto Maru* was part of a high-speed convoy that left Manila August 27, 1944, with 1,035 POWs, only one of whom died during the voyage. The hold was hot, but unlike winter ships, pneumonia wasn't a problem. The *Noto Maru* was not damaged in an attack on the convoy and reached Moji on September 4, nine days after leaving Manila. "All things being relative, the trip on the *Noto Maru* was one of the better hellship voyages," Michno wrote.[195]

The POWs boarding the *Oryoku Maru* assumed guerilla networks would get word to U.S. commanders in Australia, who would protect the ship from an attack by Americans. They were partly right: "Manila was so thoroughly wired" that intelligence could reach American forces in two hours. But American ships were alerted to target a convoy in Manila Bay. Pilots had no way to know which of the unmarked ships in a convoy carried POWs.[196]

The prisoners were crowded into approximately four thousand square feet in three compartments in the hold, or approximately two and a half square feet per person. There was not enough room for most to sit down, let alone lie down, and when the hatch was closed, the temperature rose to above one hundred degrees. Wada's response to the POWs' cries about the heat was a threat to open the hatch and start shooting. "The heat, thirst, and lack of oxygen drove men mad," according to Michno. Unlike some earlier hellships that had makeshift toilets—prisoners sat on a frame attached to the edge of the deck and defecated into the ocean—on the *Oryoku Maru*, the few buckets used as toilets soon overflowed. When food and water were lowered through the hatch in buckets, only those closest could grab some of the meager rations. Thirsty men drank urine and sweat off the inside of the hull, as well as the blood of the approximately fifty prisoners who died or were murdered the first night. "We felt as if we were trapped in an oven. ... Eerie rantings of maniacs could be heard from men crazed by the heat and frustration of it all," Lawton wrote in his memoir.

And then it got worse. During the night, the ship had inched across Manila Bay, around the Bataan Peninsula, and was near Subic Bay, where Ty had loaded supplies on Christmas Day three years earlier. At 7 a.m., just as food was being lowered into the hold, American planes strafed the *Oryoku Maru* and made multiple attacks on other ships in the convoy. The pilots didn't know the *Oryoku Maru* carried prisoners, partly because the ship's identifying number had been repainted overnight. And because Japanese ships carried both POWs and war materials, it would have been illegal for Japan to mark the ship as a hospital ship. Japanese passengers on the *Oryoku Maru*, some of whom were wounded, were taken ashore after the attack, while the POWs spent another night on the damaged ship. The next morning American planes returned, this time hitting the

Oryoku Maru with at least six rockets and one five-hundred-pound bomb. The crippled ship dropped anchor. The rest of the convoy "disappeared."[197]

On that morning, pilots could see Americans on the deck being forced to jump into the water and swim three hundred yards to shore, sometimes under gunfire. The planes wagged their wings and flew off. American planes returned several days later to bomb the *Oryoku Maru*, which sank with bodies onboard. The death toll was 286, including Ty's friend from pilot training Leif R. Kloster.[198]

Of the thousands of POW deaths on hellships, more than 90 percent were caused by friendly fire. American officials were reading Japan's detailed coded messages about convoys and ships, and therefore knew at least some of their targets carried prisoners, but that information was not made public until the late 1970s. "Perhaps the answer lies in the just indignation of thousands of surviving kinfolk who would not take kindly to knowing that the lives of their lost loved ones were played with so cavalierly. Friendly fire was not so friendly, nor inadvertent," Michno wrote.[199]

For the next few days, Ty and the approximately thirteen hundred POWs who were still alive sat, knees bent, fifty-two men across in twenty-six rows on a tennis court that was part of the old U.S. base at Olongapo, where Ty had rescued supplies from Fort Wint on Christmas Day 1941. Some of the men were optimistic, including medic Estel Myers: "We're still on Philippine soil and now we know for sure the Yanks are on their way. It won't be long now." There was a single water faucet and no drugs to treat dysentery or malaria. It was sunny and hot during the day and chilly at night. The soldiers were provided with shirts, which helped ward off sunburn. On December 17 and 18, the POWs got small servings of uncooked rice. On December 20 and 21, they were loaded onto trucks and taken to San Fernando. This was the same place where,

in April 1942, most had boarded boxcars as part of the Death
March, and Ty had slipped away. Now in 1944, the prisoners
were allowed to choose the fifteen sickest men among them to
return to Bilibid. Instead, the soldiers were taken to a cemetery
and beheaded or shot.[200]

On Christmas Eve 1944, the day after Ty's 25th birthday,
they boarded boxcars once again and were taken to Lingayen
Gulf. They arrived there two weeks before 818 ships disgorged
175,000 soldiers of U.S. Sixth Army in the same spot, January
9, 1945. The POWs foraged for plants and flowers to eat—
some got rice balls, some got nothing from their guards—and
begged for water. Twenty men shared one canteen. Later, there
was food, some of it from looted Red Cross boxes. They were
allowed to bathe in the sea. Three days later, the prisoners
were loaded onto two ships: 1,070 on the *Enoura Maru* and 236
on the *Brazil Maru*. By December 31, they were near Formosa,
where it was much colder. The POW population was down to
1,184, and the prisoners were all moved to the *Enoura Maru*
for the rest of the journey.[201]

The last word Hans and Shy received from someone who
saw Ty alive came from a soldier who had been in his 41-C
training class at Randolph and was on the *Oryoku Maru*. Lieutenant Lycurgus W. Johnson, of Boulder, Colorado, wrote about
their imprisonment and the trip on the *Oryoku Maru* in a letter
dated December 9, 1945, after he got out of a Colorado hospital. He called Ty "Koke," a nickname applied to many Kokjer
family members, and described their situation more starkly
than Ty did: "We were starved even before we left Bilibid and
our food consisted of one cup of rice for three men and one
cup of water for 4 men per day. Conditions were the worst. We
lived like rats. ... I don't know how any of us survived."[202]

On a cold day in Formosa, Johnson and many POWs were
on the deck, where at least there was fresh air. Ty and some
others were sent into the ship's forward hold. The morning

of January 9, 1945, U.S. planes bombed the *Enoura Maru* and a Japanese destroyer next to it, killing 250 to 300 POWs, including Ty's friend Alvan Ose, of the 27th Bomb Group. "The carnage was indescribable. More than half of the five hundred men in the hold were killed outright. For the wounded, shrieking in pain, there was no medicine, no dressings. Nor was there any answer from topside to pleas for help," Toland wrote. Ty's friend, Johnson wrote, "I never was so glad to see anyone when I saw Koke coming out. About seven men came out alive out of the 250 in that hold ... The Japs left them in there with the dead and dying for 4 days before they would let anything or anyone out."[203]

The surviving prisoners were moved to the *Brazil Maru*, which sailed January 13. Although the ship carried plenty of food and water, little was given to the POWs, all of whom by then had dysentery. Some bartered with guards for food and water with their last possessions, including West Point or wedding rings. It was cold, and some came down with pneumonia. Others gathered snow that fell through the hatch for drinking water. Estel Myers, whose papers were a main source for *Belly of the Beast*, described his fellow hellship prisoners: "They looked more like feral beasts than men. Their four weeks' growth of beard and their hair was matted with one another's excrement; their skin was drawn tightly over their bones; their eyes were hollow and glassy." Their rations were half a canteen cup of rice and barley soup and four teaspoons of water per day. Of the 1,619 who had begun the journey, between 425 and 450 prisoners were alive when the ship docked at Moji on Kyushu, Japan, on January 30, 1945. In the Philippines, another 1,300 American members of the military remained to be rescued.[204]

Total travel time for those who boarded the *Oryoku Maru* December 13, 1944, was 21 days. They spent nearly as much time on land as they did at sea, after the *Oryoku Maru* was bombed in Subic Bay and after the *Enoura Maru* was bombed in Takao.[205]

A Japanese order issued as the first hellships completed their journeys in 1942 complained about conditions on the transport ships. Prisoners needed to be able to work, it said. POWs who were ill or who had died were of no use to Japan. The order was rarely, if ever, followed. When the *Brazil Maru* docked in January 1945, local officials who came on board "were stunned as they took in the deck scene before them. ... Debilitated POWs were rank-smelling, slop buckets full to the brim, and a stack of naked, emaciated dead men," Pearson wrote. Even after years of abuse of POWs in camps and on hellships, these officials had expected one thousand healthy prisoners ready to work. The POWs were brought naked onto the freezing deck, sprayed with disinfectant, and tested for dysentery. Some were given clothing and shoes. Children spat at them as they marched through Moji to an unheated building, where there were some blankets and coats.[206]

"We arrived at Moji, Japan, Jan. 28 [1945] and got off the boat Jan. 29," Johnson wrote in his letter to Hans and Shy. "We were all in very poor condition. I weighed 110 and my normal weight is 195. We all were sick with dysentery and malnutrition. I stayed and went with the Fukuoka group. I tried to talk Koke into coming with me but he felt as though he couldn't make it and stayed with the hospital group."[207]

Johnson didn't say whether Ty had been wounded in the bombing of the *Enoura Maru* or if he simply felt too weak to make it to the camp. It's also possible Ty had no choice. Others reported that Japanese officials decided which prisoners would go to the hospital.

POWs worked in two dozen labor camps near Fukuoka, mostly at coal mines. Conditions were cruel, but like Johnson, many of those POWs lived through the war.

Myers, a medical corpsman, noted that no medical personnel were sent to care for the sick POWs. Pearson concurred: "The hospital group of men were in such dire condition, they were never going to recover. The Japanese probably figured there was little point in wasting able-bodied men by sending them out with the doomed." By the end of February, another 161 *Oryoku Maru* POWs were dead, including 76 of 100 men in the hospital group.[208]

One of those men was Ty. He died February 12, 1945, of acute colitis, likely caused by dysentery.

Sidney Stewart wrote in *Give Us This Day* that he was told the hospital, "a dilapidated wooden structure," where he and Ty were taken, "was a typical hospital used by the Japanese to house their insane." Stewart said the one hundred men were housed in two rooms and given blankets but no medical treatment. "Hope gradually receded from the hearts of the men around me. One by one, they began to die. Some days the Japanese neglected to bring us any food at all." When the POWs cried out for food, guards sometimes dumped slop in the middle of the floor, and the men crawled over to scoop it up with their hands. Stewart felt better after he killed and ate a rat. "The days dragged into weeks. More men died." By some miracle, Stewart was barely alive in April 1945 when British prisoners of war, guarded by a Japanese soldier, carried him out and began nursing him back to health. They turned him over to Americans in a POW camp. Friends gave Hans and Shy Stewart's memoir inscribed in Ty's memory.[209]

Several survivors wrote that those who gave up hope seemed more likely to die, and Johnson's letter raised that issue about Ty. Was he injured in the *Enoura Maru* attack? Had he given up the hope that had sustained him for more than three years? Or did he think he was going to get medical treatment at the hospital? Many prisoners believed they would be treated better once they got to Japan and near its leaders. After the war, one survivor said that love didn't keep men alive; hate did. "Those that started begging for their mothers, then you'd better start digging a hole for them. Once you felt sorry for yourself, you were an absolute gone bastard," one survivor said.[210]

Johnson was bitter: "The Japs would never give us any clothes or medicine after we swam ashore from the first boat, and it snowed on us before we got to Japan. Perhaps Koke was one of the lucky ones after all. Our treatment was very bad after we got to Japan and it was there that I received my permanent disabilities. There are 290 alive out of 1618 men who started that trip."

From the time Hans and Shy learned that Ty was on a ship at the end of 1944, they heard nothing for months. They would have rejoiced with all Americans in May 1945 when the war ended in Europe, and three months later, in the Pacific. They never wrote about the use of the atomic bombs that obliterated Hiroshima and Nagasaki on August 6 and 9 that convinced Japan to surrender. "Americans greeted the news with unbridled jubilation," historian Ian Toll wrote about the bombs and the surrender. According to Toll, the government lied to the public about the targets, saying the two atomic bombs were dropped on military targets. And for soldiers, some being transferred from Europe, dreading a likely invasion of Japan—where soldiers and civilians were being trained to fight to the death—it was a huge relief. Debates about the morality of atomic bombs began immediately after their use, including among military leaders.[211]

On August 15, Japanese civilians heard the voice of Emperor Hirohito on the radio for the first time. As a living god, the emperor's public appearances were rare, and his radio speech broke all precedents. He announced the surrender and called on his subjects to lay down their weapons. Hirohito's voice had been recorded only one before, by accident, in 1928, when he was some distance from a microphone. Few heard the earlier broadcast, but nearly all Japanese citizens stood at attention to listen to the surrender. By the end of the month, Radio Tokyo was reporting that some Japanese citizens were committing suicide in protest in front of the Imperial Palace as 374 American ships sailed into Tokyo Bay. The United States began the postwar occupation of Japan on August 30, and the formal surrender aboard the USS *Missouri* took place September 2.[212]

On August 29, 1945, not knowing Ty had died six months earlier, Shy wrote in her scrapbook: "Tonight MacArthur returns to liberate the last of his men taken on Bataan, and that means you, Ty. While I write this your favorite piece is being played [on the radio] it just happened that way, it seems we cannot wait, but you and your Dad and Mom have lived to this happy day and soon we shall be together again, I sincerely hope we can repay in many ways and always for all you have suffered. Now we shall wait to hear from you and if you are delayed on the coast we can be there with you." It is the last entry in the scrapbook. Shy saved a few newspapers, including the February 5, 1945, *Morning World Herald*. The banner headline was encouraging: "Yanks Push Into Manila." A smaller story repeated MacArthur's 1942 pledge: "I shall return."

Beginning in late August 1945, Shy wrote several letters addressed to Ty in San Francisco, where most former POWs from the Pacific theater were brought by ship after stopping in Manila to get replacement papers and uniforms. Like the early war letters, they all came back. On August 27, Shy wrote from Omaha, where Hans was on a business trip. "We know you will be in Manila again. I just wonder what all your reactions have been. ... I'm sure it is a real joy coming back liberated. Time just can't go fast enough, Ty, until you get home."[213]

She added they had taken six months off after leaving the hotel, staying with her father and then Tom Kokjer's family. Before moving into their apartment in Lincoln, they were at the Cornhusker Hotel for four weeks, where they met soldiers who had returned from the Philippines.

On August 31, with the temperature in Lincoln at one hundred degrees, Hans and Shy hosted friends for a dinner of baked pork chops, hot dressing, and buttered potatoes. After dinner they played bridge in their un-air-conditioned apartment and listened "to the radio so we don't miss a word of news. The occupation is on in earnest and they must be contacting all of you now or soon, and we are so anxious, Ty, it is really hard to do anything but think of you, but we must, as we still have time to wait." In a letter the next day, they enclosed a note from Hans's sister, Meta Kokjer Key, who wrote, "Hurrah! The War is over! The first thing that came to my mind was Ty."

Four days later Shy wrote, "We look at your pictures and talk to them just as we will to you." She reported that Hans had sprained his ankle and they had seen Ty's cousins Margaret Cobb and Phyllis Kokjer, who were in Lincoln to go through sorority rush week. Shy repeated her expectations they would see him soon. "Well dear Ty we don't get far from thinking of you, you, you. Time is passing and soon we shall be hearing. Today they signed the peace and surely you will be found and moved real soon." They would fly to San Francisco if that would speed their reunion.

The following week there was more news about Margaret and Phyllis and life on the University of Nebraska campus. "Ty you have to hurry back. It seems there are so many girls they are flooding the campuses. Well, anyway, you just get back, nothing really matters but you and seeing you again soon. Some of the boys who were with you at Cabanatuan talked from Manila today. Wonder if they knew you."

Shy's last letter to Ty was dated September 12 and postmarked the next day. They had received documents from the Army Effects Bureau, which included "some personal papers, marked as belonging to your son, which have been damaged by water." She wrote they would have the letters (the diary) typed so they could share them. "We had a real thrill yesterday. Some of the letters you had written when you were in hiding from the Japs, were sent to us, and also three of our letters sent to you. And a few little paper clippings. ... They are so dear to us. ... Time can't go fast enough. We want you home so badly we can hardly wait. But when we get you, you won't [be] able to get far away from us."

The War Department "report of death" document says Ty was in "beleaguered status," which means at war, from December 8, 1941, to May 7, 1942, when he was declared missing in action. His total time in service was three years, four months, and seventeen days. The adjutant general's office received on September 12, 1945, "evidence considered sufficient to establish the fact of death" from the provost marshal general, which had received information through the Red Cross. Ty's obituary ran on the front page of the Lincoln newspapers September 17, 1945. Some friends and relatives first heard of his death on radio newscasts.

The family files include nearly one hundred sympathy cards and letters from friends, cousins, uncles, and aunts, including one from Meta. She said people in Clarks were "just crushed," adding "I can't seem to collect my thoughts anymore, but you are on my mind continually. It's bad enough to have to give Ty up after all this time, but the way he had to go really hurts." U.S. Cobb wrote that he knew Ty had died bravely. "I hardly feel it's right for these boys to die for mistakes they never made. I must not feel bitter so will keep my thoughts to myself." Emerson wrote, "I wish there were some way I could ease your sorrow, but all I can do is share it with you." Carter Kokjer, Tom and Isabel's oldest son, wrote from a Navy training school at Notre Dame University, "I have known him as long as I can remember, and now that he is gone, I wish I could have known him better. He was always swell to me and I enjoyed being with him. I am proud that I knew him, because, in my eyes, he was one of the many brave men who survived the worst that the war could offer. ... It's up to me and the rest of Ty's generation to keep anything like this from happening again and we shall."

Bobby Cobb wrote from his home in Houston, inviting Hans and Shy to visit. "It isn't as if I've just lost a cousin, because Ty has been a great deal more than that, and he's been a great deal more than a 'pal' too. Ty was closer to being a brother of mine, I guess—not because of blood relationships but because he was so doggone swell, and we shared so much in mind and spirit." Jack Obbink, who had been a POW with Ty, sent this message: "It is with the deepest regret I have ever felt that I write this letter and there is so very little that I can say. ... I feel that I have never known a truer friend than Ty. There are few men who can say they have done their duty to God and their fellow men as he did." Judge Messmore, whose son survived, wrote, "Our family was deeply grieved that your son, Lt. Madsen Kokjer, paid the supreme sacrifice after going through so much that most of us felt assured would result in his liberation."

Hans and Shy also heard from government officials. A letter from Army Air Forces Commanding General Henry A. Arnold read in part: "Determined to uphold the high traditions of this organization, he performed the exacting duties of an officer in a worthy fashion. His fine character and friendly disposition made him well liked by those with whom he served." MacArthur praised Ty as "gallant" in a condolence letter: "You may have some consolation that he, along with his comrades-at-arms who died on Bataan and Corregidor and in prison camps, gave his life for his country. It was largely their magnificent courage and sacrifices which stopped the enemy in the Philippines and gave us the time to arm ourselves for our return to the Philippines and the final defeat of Japan. Their names will be enshrined in our country's glory forever." A card from President Harry Truman reads: "In grateful memory of Lieutenant Madsen C. Kokjer / Who died in service to his country."

Sometimes as battles still raged, or immediately after fighting ended, soldiers and the American Graves Registration Service began the long task of recovering the remains of soldiers from hundreds of temporary graves across battlefields, and reburied them in cemeteries abroad and at home. Dog tags were found on some bodies, but many remained unidentified. Japanese soldiers took some dog tags as souvenirs, others were lost, and at some camps POWs gathered dog tags as a record of a soldier's death. In some places, remains were commingled in mass graves that families and the military are still trying to sort out, now aided by DNA. But Ty's ashes were preserved intact, and the Army wrote to Hans and Shy that they could choose his burial site. There is no written record about how they decided where they wanted Ty's ashes to be buried. They are both buried in the Fridhem Cemetery south of Funk, where Shy's parents are buried. The National Cemetery of the Pacific in Hawaii or Arlington National Cemetery would have been among their choices.[214]

The American Battle Monuments Commission maintains twenty-six military cemeteries on foreign soil. The most-visited U.S. overseas military cemetery is Normandy American Cemetery in France, which includes the graves of 9,380 American soldiers, most of whom died on D-Day, and lists the names of 1,557 who are missing. The Philippines are home to another, less famous cemetery.

Manila American Cemetery, south of Manila, is the former site of the U.S. Army headquarters at Fort McKinley (later renamed Fort Bonifacio). Ty worked there during the early days of his deployment, and it is his final resting place.[215]

In February 1958, Hans and Shy visited Japan and the Philippines. Photos from their visit show them with members of the Jose family standing next to the cross that marks Ty's grave. Madsen C. Kokjer, First Lieutenant U.S. Army Air Forces, 17th Bomber Squadron, 27th Bomber Group, Light, is buried in Plot B, Row 11, Grave 121.

First Lieutenant Madsen C. Kokjer,
17th Squadron, 27th Bombardment
Group, Light, was one of 400,000
Americans in the military who died in
World War II. He is buried at Manila
American Cemetery, which includes
the largest number of overseas World
War II graves. Another 36,286 soldiers
who are still missing in action are
commemorated on plaques. Photo
taken in April 2023 by Jody Beck

Acknowledgments

No one writes a book without help, and I am grateful for the relatives, friends, and people I've never met for guidance and information.

Hans and Shy Kokjer held onto Ty's papers and had the decaying, hand-written diary transcribed. After they died, my aunt Janet Kokjer Sothan and her husband, Norm, carefully stored the diary and hundreds of letters for more than 40 years, hosted my visits when I copied the documents, and answered many questions. This book could not have happened without them.

I owe many thanks to those who thoughtfully critiqued proposals, read the manuscript, and gave invaluable advice. Blenda Femenias read and edited the entire manuscript, which meant it was in much better shape when I submitted it for a final edit to Katherine Pickett at POP Editorial Services, whose edits improved the storytelling and style. Others read proposals and parts of the book and always had something useful to say: Karl Kellar, Mindy Greiling, Charles and Mary Stewart, Katherine Skiba, Marc Jacob, and William H. Hammond. Carol High recruited a friend to translate Japanese phrases. To my friends in the Independent Authors, an informal group of writers at the National Press Club, thanks for advice to a rookie in the publishing world. And to my fellow journalists and writers of JAWS (Journalism and Women Symposium) for seminars and personal advice.

Gary Boyd, command historian, Headquarters Air Education and Training Command, Joint Base San Antonio-Randolph, Texas, hosted me at Randolph Field, for several days of research. He read the manuscript, and has continued to provide information. Reagan Grau, director of collections and exhibits, and Robert Elder, collections registrar, at the National Museum of the Pacific War, in Fredericksburg, Texas, arranged for me to visit and photograph World War II canteens. Lisa Parish, at the Kearney Public School Foundation, found Ty's senior high school yearbook and allowed me to take photographs. Jennifer Holthus at the Grant County Library guided me in an afternoon of research and ended up joining me in my quest for information about the Kokjers' time in town. K.C. Schaack took time from a busy day to meet with me to talk about the Hotel Hyannis, which her family ran until recently, when they sold it to a longtime employee. She took me to the third floor, where Hans, Shy, and Ty lived, which wasn't in use in 2017. Ginger Fouse who runs the Grant County History Museum, opened the museum to allow me to visit on a day it otherwise would have been closed. Kaia Wahmanholm, at the National Nordic Museum, is one of many curators who quickly answered questions.

Veterans and their families were always willing to share stories. I am particularly grateful to Dan Crowley, a Bataan Death March survivor, who died in the summer of 2021 at age 99. Someone thought he was in Ty's unit. He wasn't, but he still talked to me about his war experiences. His wife, Kelley, was part of the tour I took to the Philippines in 2023. I had to substitute a phone call for an in-person meeting with Joe Wilson, a retired Army colonel, when I could not travel to San Antonio in 2020. His father Ovid O. Wilson (known as Zero Wilson) was on the Death March when Joe was a child. He shared memories of his father and his family's worrying time at home. Darcy Pattison, a children's book author, and her sister, Elleen

Hutcheson, told me about their father Henry Thomas Foster's experience on the Death March, at a POW Camp, and on a hell-ship. After the war, he extended kindness and a sympathetic ear to veterans as they coped with post-war stress, particularly to one man who saved his life. Sancho Sisson and his mother, Rose, told me about her family's courage in helping guerillas at her childhood home on Luzon Island, even when Japanese soldiers were on their property.

Family members and friends have heard me talk of little else the past six years. Cousins Anne Potter and her husband Paul, hosted me over many nights in Lincoln, and Ryan Sothan lent an ear. (Anne and Ryan are two of Janet's children. Her third child, Laura Spendolini, lives on the West Coast.). Diane and Tom Kokjer hosted me at their home in Sidney, Nebraska. My second cousins, the children of Ty's cousins—Carter Kokjer (son of Tom and Diane) and Emerson Scott (son of Ann Kokjer Scott)—and second cousin once-removed, Tom Kelly (great-grandson of Meta Kokjer Key) provided information and support. Several friends have provided morale boosts and have given me books or suggested resources, including Carol Guensburg, Carolyn Cerbin, and Sandra McElwaine. Friends in Washington and my many friends from Lincoln always asked how the book was going and promised to help me promote it. To those friends I haven't named, I also owe you thanks.

Thanks also go to my siblings and their families, who hosted me, listened to me, and provided enthusiastic support: Marcia Brown and Bill, Jeannine Keairnes and Scot, Larry Beck, and Ron Beck, who died in October 2021, and his wife, Betty.

I was taught that it's a good idea for writers to have an audience in mind. For this book, I thought about the generation of my nieces and nephews. Some of them studied World War II in high school or college and knew about the Bataan Death March, and others didn't. I hope they and other members of their generation will find the tale of their grandmother's cousin interesting, and I owe all of them a debt for listening to me: Jon Pace Beck (Lizzie Pace), Alex Beck (Ashley Slanek), Alida Beck; Sam Keairnes (Zelina), Peter Keairnes (Abigail Schmidt); Alyssa Maddux (Rob), Laura Hays (Tyler), Carly Brown, and Joseph Brown (Molly), who earned special thanks for spending a rare day off to photograph many of the documents and photographs in this book and my portrait.

After considering several options, Janet decided to donate Ty's papers to the Veterans History Project at the Library of Congress. Megan Harris and the rest of the staff there have been welcoming on the occasions I have returned to work with the papers, which I did not want to risk keeping in poor storage conditions at my home. Elsewhere at the Library of Congress, librarians were always willing to track down an old magazine or newspaper and advise me about my research. Staff at the Smithsonian Institution's Archives of American Art were also helpful. James Zobel, MacArthur Memorial archivist, provided information and kept me in the loop for online talks by authors and experts.

I have been able to talk to people whose parents, grandparents, or other relatives of veterans of the war in the Philippines through the American Defenders of Bataan and Corregidor Memorial Society. Although two in-person conferences had to be canceled or moved online, the community is always willing to provide contacts or information via email. To others doing research about relatives who were in the Philippines, I recommend joining the group. Similarly, Holly Rotundi, executive director of Friends of the National World War II Memorial, and her staff have provided information to me and excellent seminars about the war. Cecilia Gaerlan, executive director of the Bataan Legacy Historical Society, based in San Francisco, puts on seminars and has created a curriculum about World War II in the Philippines.

When I was in high school in the late 1960s, World War II was so recent, it was tacked on at the back of our history books. We rarely made it that far. We were more interested in our generation's war in Vietnam. My early knowledge about World War II came from watching 1950s movies about the war. "The Longest Day," starring John Wayne and a dozen other famous actors, tells the story of D-Day. "Sink the Bismarck!" is about the British navy sinking Germany's largest, most powerful battleship. We also watched "Hogan's Heroes" a popular TV comedy, depicting American POWs outwitting their German POW camp commander. "McHale's Navy," another TV comedy, depicted Americans stranded on a South Pacific island who lived a pleasant life until an officer parachuted in to bring military order to the reluctant sailors. "The Bridge on the River Kwai," set in Burma and starring William Holden and Alec Guinness, was more realistic. It showed the cruelty of Japanese guards toward some POWs. I remember my strong reaction to the movie and its final scenes. But I don't recall watching any of the few movies made about Bataan or the Philippines.

Later in life, I began to read histories and biographies, mostly about the 20th century. That knowledge and my discovery of Ty's story when I was an adult set me on the adventure that became this book.

Epilogue

"It was close, but that's the way it is in war. You win or lose, live or die—and the difference is just an eyelash."

—General Douglas MacArthur, after arriving in Australia, March 1942

We all make choices that affect the rest of our lives. In Ty's short life, he made many. Some led to the joy of flying solo above the clouds. Or the possibility of a future in the country with a girl named Mary, in a house he planned during the months he hid from Japanese soldiers. Those decisions also set him on the course that led to his death in Japan. As in all wars, the decisions of others, some made by those close to him, others made at the highest levels, had much to do with Ty's fate. In late July 1942, Ty wrote about a grave situation in Guagua, where guerillas were hoping to kill the pro-Japanese mayor. He had written a letter to his hometown friend and fellow pilot Jim DeWolf, who was driving the night Ty's leg was broken. Ty hoped he could send the letter some day and that DeWolf would receive it. But he wasn't confident: "This war creates lots of ifs."[216]

What if the peace negotiations between Japan and the United States in 1941 had turned out differently? If Japan had withdrawn from Indochina and agreed to other demands from the United States, and if Roosevelt had eased up on oil shipments instead of blocking them, war might have been averted or delayed.[217]

What if Ty had stayed in college and waited to be drafted or had enlisted later? What if he had deployed with his original pilot-training class? What if the Army had sent him to Honolulu, or Panama, or assigned him to fly transport planes instead of light bombers? Several of Ty's friends, including DeWolf, were sent to Pearl Harbor, flew combat missions, and survived the war.

If the 27th Bombardment Group's planes had arrived in the Philippines before the war started, Ty could have died on a bombing mission, or he could have escaped. His friend Hiram Messmore deployed months earlier than Ty and was one of the last pilots to fly out of the Philippines to Australia. He flew in several battles and survived the war.

What if Ty had decided to be an active guerilla, gathering information, harassing Japanese soldiers, then disappearing in the mountainous jungles of Bataan? What if he had been able to hide beyond December 1942? What if he had gone with his friend Robert G. Bjoring, who decided not to surrender in April 1942? Bjoring and three others spent months moving south, hoping to escape to Australia. After being recaptured, Bjoring spent time in Bilibid Prison and was sent to Japan later that year. He survived the war and wrote a memoir that includes his friendship with Ty.[218]

If Ty's health had declined sufficiently at the Cabanatuan prisoner-of-war camp, he might have been among the five hundred POWs who were rescued in January 1945. They were some of the first soldiers who returned to the United States from the Philippines.

In late 1944, those what-ifs became more consequential. Ty surely knew that on September 21, American bombers began attacks near Manila. It was a moment of joy. Despite the risk that those bombs might kill them, prisoners realized that

United States forces were nearby and the war was likely to end soon.

The window for Japan to transport the last prisoners in the Philippines was closing. More transports were being attacked and sunk by U.S. planes or submarines. One of them, the *Arisan Maru*, was sunk October 21, 1944, the day it left Manila for Japan. Of the 1,800 POWs on board, 1,792 died. Ty and the other 1,618 POWs bound for the *Oryoku Maru* had been at Bilibid Prison for more than two months before they were put on the doomed ship. Ty's friend Leif Kloster died when the *Oryoku Maru* was bombed in Subic Bay. After being transferred to the *Enoura Maru* and the *Brazil Maru*, the surviving POWS sailed out of Lingayen Gulf for Japan in late December, two weeks before U.S. troops landed in the same spot. What if Americans had landed earlier, or the Japanese ships had been delayed for two weeks? Ty and the others might have been killed, but they might have been left on the beach or taken back to Cabanatuan and rescued at the end of January.[219]

Had he been kept at Bilibid, Ty might have survived. The prison was one of the first buildings in Manila entered by American soldiers. MacArthur visited the prison February 7, 1945, five days before Ty died in Japan. Most of those soldiers were soon sent home. Ty left a letter with a friend at Bilibid, who survived and sent the letter to Hans and Shy when he returned to the States. At Bilibid, POWs could have been murdered by the guards, but they simply walked away as U.S. Army units neared. The slaughter of nearly one hundred prisoners at Palawan, a small Philippines island southwest of Luzon, is a cautionary tale. Telling POWs on December 14, 1944, that an air raid was imminent, Japanese guards herded the soldiers into long trenches, doused them with gasoline, set the trenches on fire, and shot into them with machine guns. Somehow, eleven POWs survived.[220]

What if Japanese forces had withdrawn from Manila and declared it an open city in January 1945? MacArthur expected the Japanese to do what he had done more than three years earlier to save the city and protect civilians. Instead, Japanese soldiers destroyed blocks of buildings and murdered as many as one hundred thousand civilians as they fought to their own deaths or committed suicide.[221]

What if, at the end in Japan, Ty had gone to the coal-mining POW camp as his friend urged him to do, avoiding the so-called hospital? The friend survived and wrote to Ty's parents after the war, recounting their journey from Manila. Of course, Ty could have died at the camp or on his way there. Throughout the war, Japanese doctors refused to treat POWs or provide drugs. Captured American military doctors and medical corpsmen had few resources for treating soldiers, although they sometimes worked miracles. At the so-called hospital in Moji, Japan, where Ty died, prisoners had access to water, but sanitation was poor, and food was often scraps left over from guards' meals.

What might have Ty done had he survived the war? It would have been nice to have known him as a real person, not just as a face in a few photos and through his writing. A 1936 comical gossip column in his high school newspaper reported: "His habits are to eat three times a day, go to bed at night, get up in the morning, and be friendly to everyone." His friends described him as happy, friendly, and hardworking. He described himself that way, too, but also confided occasional doubts and sadness.

Could he have rekindled his romance with Mary Tice? She apparently led a happy life. Mary Tice Schweitzer died in 2010 at age 89. She lived in the Chicago suburb of Northbrook, Illinois. She and her husband, Jack, who died in 1998, had four children and two grandchildren.

Millions of Americans fought overseas or moved away from their hometowns to work during the war. Those "soldiers and sailors finally got to see the world, and so formed a broader, more enlightened picture for themselves. Many never returned to their home places afterward, thus resulting in a significant shift in the U.S. population," Groom wrote.[222]

Ty wrote plenty about what he wanted to do when he got home, and some of it reflected a broader world view. He acknowledged that the war had made him a more serious and tolerant person. Memoirs by other veterans make the same points. Ty had planned the house he wanted to build in Hyannis, down to the furniture and where he would use carpet or linoleum on the floors. Helping his parents run the hotel, which is still in business under local ownership, and perhaps adding another one nearby, was always among the possibilities. His alternate plan was to marry and live elsewhere. He had no way of knowing that Hans and Shy had given up the hotel and were living in Lincoln by the time the war was over. After a few years with the state government, Hans became a company executive for a short time.

The Hotel Hyannis serves a different clientele than the cattle buyers, traveling salesmen, dentists, and opticians who stayed or held office hours there in the 1930s. When I stayed there in the summer of 2017, the rooms on the second floor were filled, mostly with members of a road crew who were repaving Highway 2 east of town. The third floor was no longer in use, although the owners said they would fix it up if they thought it would pay off. Town and county residents were more likely to stop by the hotel restaurant for lunch than for dinner. The bar closed at 9 p.m. on weeknights.[223]

Ty's parents lived in Funk for a few months after Shy's father died in December 1947. They briefly ran a hotel in Holdrege, which is southwest of Kearney and six miles west of Funk. They called it The Madsen in Ty's memory. A story published

January 8, 1947, in *The Holdrege Daily Citizen* praised the hotel, saying Shy had shown "remarkably good taste" in redecorating and that the couple were sincere and friendly. The hotel is now an apartment building.

By the 1950s, Hans and Shy were living in a garden apartment in Holdrege, where the population has hovered near five thousand. When they retired, they continued to visit relatives, including an annual summer birthday gathering—Isabel, Shy, and Winifred all had mid-July birthdays—at one another's homes. Hans and Shy took cruises with friends, including one to Cuba, and visited Lincoln, where my family lived; Wahoo, where Emerson and Winifred lived; Clarks, where Hans's sister, Meta Kokjer Key, lived; Larkspur, California, where Roy and his family lived; and Sidney, where Tom and Isabel lived. Audio tapes recorded in 1967 by Emerson, Hans, and Shy recount Emerson's visit to Funk and Holdrege and a busy weekend in Wahoo and Lincoln centered on a University of Nebraska football game. Had Ty survived, he surely would have been part of many of the weekend events.

Some of Ty's pals stayed in the Army and became U.S. Air Force pilots, once the service became independent of the Army in 1947. That was something Ty had considered, along with becoming a commercial airline pilot. Ty's pilot-training class 41-G held annual reunions for many years. There were occasional newsletters and reports about those who had died. Pictures in the newsletters show happy comrades at dinners or cocktail parties with their wives, or posing in group photos that included fewer people as the years went by.

Sixteen million Americans served in the military during World War II. As of 2022, just over one hundred sixty-seven thousand remained alive. A year later, that number was one hundred nineteen thousand five hundred. For some of those imprisoned across the Pacific theater, coming home was hard. Four years away, most of that time in captivity that was truly cruel, made adjusting difficult. They were among the first to go down to defeat in the war, and many felt like losers among other soldiers who had fought more recent, victorious battles. Some had physical disabilities for the rest of their lives, and they were likely to die at a younger age than other veterans. Others had what we now know as post-traumatic stress disorder, or PTSD. They ducked at loud noises, they hoarded food and other supplies, they overate at buffets, they had a hard time trusting people, and sometimes they had visions of dead comrades walking down the street. Their girlfriends had married while they were overseas. Some wives, assuming their husbands were dead, had remarried. Their children didn't recognize them. Virtually no one wanted to hear about their lives as POWs. One veteran, who nearly starved to death as a prisoner, was enraged when a friend complained about how hard things had been at home because of gasoline rationing. Those who felt the need to talk about the war sometimes alienated friends and relatives.[224]

The former POWs campaigned for fifty years for compensation that, when it came, they viewed as too little too late. Another thorn was that Japanese American civilians who had been held in internment camps in the United States during World War II were given $20,000 in restitution in 1998, following a decades-long campaign. The fifty-three American diplomats held hostage by Iran for more than a year were awarded as much as $4.4 million under a law passed thirty-six years after their capture. Meanwhile, the U.S. government allotted soldiers held by Japan just $2.50 for each day they were held as POWs and subjected to inhumane treatment or forced into hard labor. Under those rules soldiers who were POWs from April 9, 1942, to February 7, 1945, would have been paid about $2,360. POW lawsuits against Japan for compensation were thrown out because of provisions in the peace treaty.[225]

Even today, receiving recognition has been difficult. In 2015, a law awarded the Congressional Gold Medal to Filipino veterans of World War II. There is no comparable medal for American veterans, although efforts to create one continue.

More civilians than members of the military died during the war, which might surprise some. Approximately 22 million combatants died in World War II. Fifty-five million civilians died. Westminster Abbey in Great Britain maintains a list of the names of civilians who died in that country. The National World War II Museum has a list of the number of civilian deaths by country. Some other countries memorialize civilian deaths, but often without names.

Most American soldiers returned seamlessly to their civilian lives, slightly interrupted. By 1951, more than 8 million veterans had used the G.I. Bill to finish high school, learn skills, or graduate from college. The end of the war ushered in a time of prosperity. Factories that had produced goods for the war became peacetime factories where veterans got jobs satisfying pent-up needs and wants for things that weren't available during the war years, including cars, washing machines, and television sets. (In some instances, veterans displaced female workers, some of whom wanted to keep their jobs.) Veterans bought those cars, got married, birthed the 76 million children of the baby boom generation, and used G.I. Bill–backed loans to buy suburban houses for their large families.

That was certainly true of my family. Leo J. Beck Jr., my father, who was four years younger than Ty (they never met), served in the Army. A double bout of rheumatic fever, which caused heart and kidney damage, meant he never went overseas or saw combat. He trained other soldiers and was medically discharged in late 1944. He returned to the University of Nebraska, where he had completed one semester, graduated in 1948 with a business degree, and became an insurance broker in Lincoln, where he grew up.

Phyllis Kokjer Beck, my mother, six years younger than Ty, worked at an ordnance factory in Mead, Nebraska, about half-way between Wahoo and Omaha, on her breaks from school during the war. She and the 3 million other women who worked at ordnance plants all over the country were called WOWs— women ordnance workers. Mom was slender and crawled into bomb casings to install fuses and drove an electric truck. She graduated from a two-year college in Wahoo, then attended Nebraska, graduating in 1947. Although she was two years younger than Dad, his wartime service put him behind her in school. They married a week after he graduated, June 12, 1948. By the end of 1961, my family included five children. We had occasional visits from Hans and Shy. She bubbled about family news, and Hans, who was quieter, would come up with quips or jokes.[226]

Could that life have happened for Ty? Could the outgoing man who whistled his favorite music, saw all the movies that came out, made friends easily, and had a vision for his postwar life actually lived it? Some of his friends did. Bob Bjoring married the woman he met in San Antonio during pilot training and remained in the Air Force, mainly managing base grocery stores in the United States and abroad. One postwar assignment took him back to the Philippines, where he met with some of the people who helped him escape from the Death March. After his military retirement as a lieutenant colonel, he worked as a manager in several California cities, then retired in San Antonio. After his wife died, he married the widow of another war veteran. Following his death in 2006, his wife donated his papers to the Headquarters Air Education and Training Command, Joint Base San Antonio–Randolph, Texas.

Jim DeWolf, Ty's friend from the second grade through pilot training, flew B-17 and B-24 bombers in forty-one combat missions in the South and Central Pacific, including in the battles of Guadalcanal and the Bismarck Sea, off Papua, New Guinea.

He was awarded the Distinguished Flying Cross, the Air Medal, and the Soldiers Medal. DeWolf remained in the Air Force for thirty-four years and rose to the rank of colonel. He and his wife had one son. They retired to Edinburg, Texas, where he died in 1995 at age 75.[227]

Hiram Messmore married an Army nurse he met in Australia, was promoted to major, and by 1943 was stationed in the United States. He attended law school, returned to Nebraska, and died in 1993.

Tom Ward, one of Ty's Pampanga roommates, wrote to Hans and Shy after the war hoping to reconnect. Merle Musselman practiced and taught medicine in Omaha.

Three of Ty's friends in the 27th Bomb Group died during the *Oryoku Maru* journey. Alvan Stanley Ose died December 14, 1944, when the *Oryoku Maru* was bombed in Subic Bay. Ose was awarded the Purple Heart. His body was never recovered. He is memorialized on a plaque at the Manila American Cemetery and at the Story City, Iowa, cemetery. There is conflicting information about the date of death of Leif R. Kloster, of Eagle Grove, Iowa. Some references list his date of death as December 14, 1944, but the date on the tombstone at his hometown cemetery lists his date of death as January 27, 1945, the day before the ship arrived in Moji, Japan. George S. Davis died January 9, 1945, the day the *Enoura Maru* was bombed. His remains may be among those in a mass grave at the National Memorial Cemetery of the Pacific in Honolulu, Hawaii. He is memorialized on a tablet at the Manila American Cemetery that lists him as missing in action.[228]

Like many states, Nebraska has become more urban over the years. Kearney has more than tripled in population, to 34,000, but Hyannis has shrunk to less than half of its prewar population, from 449 to 162 in 2020. Omaha's and Lincoln's populations exploded as rural areas emptied. With his parents in Holdrege, Ty might have decided to return to Kearney. Or as so many people did after the war, he might have moved to Lincoln or Omaha, where the jobs were.[229]

In books and articles, there are many descriptions of the beauty of Manila and the wild jungles of Bataan before and during the war. Covid-19 postponed my trip to the Philippines three times—in 2020, 2021, and 2022. I finally made the trip in April 2023. Now I can describe some of those places myself, although much has changed in 80 years. I toured Manila, Subic Bay, Corregidor, Bataan, the Cabanatuan POW camp memorial, and Bilibid prison. I laid a wreath at Ty's grave at the Manila American Cemetery. My emotions in those moments caught me by surprise, given that I never met Ty. But I felt in those moments that I knew him, and I was tracing his steps.

Covid finally caught me at the end of the formal tour, meaning I wasn't able to visit Pampanga and the descendants of families who helped Ty. Instead, Guill Ramos has made two trips to do interviews and take photographs. The elderly children of Dr. Jose Jose recall their own terror when Japanese soldiers knocked on the door looking for American soldiers. They said their father was a generous man, and he likely gave cash or supplies to American soldiers in hiding. However, they had no specific memories of Ty or other individual soldiers.

The World War II Memorial in Washington, D.C., honors the 16 million Americans who served in the armed forces during that era. Four thousand gold stars represent the what-ifs and the bravery of the 405,399 members of the military who died for their country. I have a framed photograph of those stars, and every time I look at it, I think of Ty, the what-ifs of his life, and of those who knew and loved him.

Family of Madsen "Ty" Cobb Kokjer

Hans Madsen Kokjer (February 29, 1857-November 20,1932) was Ty's paternal grandfather. He immigrated to the United States from Denmark in 1874, at age 17. He was the third of 11 children and the oldest son. He saw no future for himself in Denmark, which lost territory after a war with Germany when he was a child. Denmark's population was growing, limiting opportunities for young men like him.

Hans watched as his father, Soren Madsen Kokjer, worked for a large landowner, then came home to work their smaller farm. When the boys were old enough, they herded livestock instead of going to school. At night when the family was together, they sang as they worked, making rope, candles, or clothing. Hans was one of 300,000 Danes who immigrated to the United States between 1840 and 1920. During that time, more than a third of the growing population of five Nordic countries emigrated. Hans brought the family's work ethic and love of music with him to the United States, and he made sure all of five of his children, including his only daughter, went to college.[230]

It was customary in Denmark in the mid-19th century for families to adopt as a surname the region where they lived. Ty's great-great-grandfather was Mads Madsen, but because the family was living in the Kokaer region of Denmark outside Kolding, his great-grandfather's name became Soren Madsen

Kokaer. The name was Americanized to Kokjer when Hans immigrated to the United States. Any American you meet who is a Kokjer, or descended from one, is related to Ty and to my family. The name is pronounced COKE-yer. The K or Ko plus joer (and other spellings) means cow plus marsh, fen, or bog—the place where cows go to drink.[231]

Hans spent two years near Charlotte, Iowa, about 16 miles west of the Mississippi River. He worked as a farmhand for the family of a Danish cousin and was apprenticed to a black-smith, where he became a master mechanic and wagon maker. He moved to Clarksville, Nebraska, in 1879. The town, soon renamed Clarks, is near the Platte River 111 miles west of Omaha. Hans worked on a farm, then opened a blacksmith and farm implement shop. He played in the Clarks Cornet Band and served on the town board. In 2020, the village's population was 344, down from a high of 605 in 1910. Clarks' most famous native is Evan Clark Williams, the founder of Twitter.

Sarah Malina Alice Hartwell (August 28, 1857–February 18, 1920) married Hans on June 26, 1883. Her family was the first to settle in Clarks, where her parents and the children ran Junction Ranch, which served stagecoaches and other travelers. The trading post sold goods to new settlers and members of the Pawnee Tribe who lived nearby.

One of Hans' sisters in Denmark, known as the best seam-stress in Kolding, made Malina's wedding dress, which was carried to the United States by his brothers, Mads, who joined Hans' business, and Jordan, who became a Congregational minister. The dress is at the Nebraska Historical Society and was included in exhibits about the evolution of wedding fashion and plains life. Mads wrote a short article about the wedding for the *Clarks Weekly Messenger*, noting that the small group was still too large to sit at the table in the Hartwell home, so they ate cake and ice cream from plates on their laps. Ice cream must have been unknown in Denmark, because Mads described it as "cream with sugar which is frozen then it is to eat like snow."

Hans eventually sold the business to Mads and became a regional salesman for a tractor company. He was active in the Nebraska State Fair committee for many years. One year he traveled for several weeks throughout the Midwest on a train outfitted as a promotion for Clarks and Merrick County. Rail cars were transformed into museum spaces, exhibiting the area's industry and economy. Letters to his family in Denmark (the original Danish and the English translations) are at the Nebraska Historical Society. After Malina died, Hans visited his family in Denmark, his only return trip to his native country.

Hans and Malina had five children:

Meta Elizabeth Kokjer Key. (January 18, 1884-March 16, 1980). She lived in Clarks except for when she attended Oberlin College and Conservatory in Ohio and the University of Nebraska in Lincoln. She gave piano lessons and played the organ at church. Her husband, John W. Key (July 5, 1882-March 26, 1936), was a farmer. They had two children.

Ralph Leroy Kokjer. (June 4, 1886-1980). He was sometimes called Ralph and sometimes called Roy. His wife was Helen Elizabeth Lovell Kokjer (September 25, 1888–April 16, 1969). He worked for the railroad, then moved to Larkspur and later to Vallejo, California, near San Francisco, where he ran a hardware store. They had three children. Ty visited them in 1940 on his way to pilot training in Oxnard. Hans, Shy, and Ty spent a week there in 1941 before Ty deployed to Manila.

Hans Madsen Kokjer Jr. (November 10, 1888-March 17, 1969) was Ty's father. Charlotte "Shy" Cobb Kokjer (July 17, 1897-June 16, 1972) was Ty's mother. Shy was born in Funk, Nebraska, and graduated from high school there. Hans attended the University of Nebraska in Lincoln. He was a steward at the state youth home, a salesman for a wholesale grocery and janitorial supply company, and worked at the tuberculosis hospital in Kearney. In 1939, the family took over the Hotel Hyannis, in Hyannis, Nebraska. In 1945, Hans and Shy moved to Lincoln, where he was an inspector for the Nebraska Department of Labor. He and Shy ran a hotel for about two years in Holdrege, Nebraska, called The Madsen. They retired in Holdrege. She was a homemaker, a volunteer aide at the tuberculosis hospital, and a full partner in running the hotels. Ty was their only child.

Thomas Edgar Kokjer (December 18, 1891-August 10, 1956) and his wife, Isabel (Izzy) Hawkins Kokjer (July 18, 1893-March 31, 1987) lived in Sidney, Nebraska, in the state's panhandle. They were close to Hans and Shy. Tom operated a car dealership and repair shop. They had three children, Carter, born in 1925; Tom, born in 1927; and Ann Kokjer Scott, born in 1933. The elder Tom was a World War I pilot. Because of injuries in a training flight crash, he did not go overseas. Isabel was one of the physiotherapists who treated him after the crash. Tom became a flight instructor at Kelly Field in Texas. Both sons joined the Navy during World War II. Carter, who became a lawyer in Kansas City, was in advanced training at Notre Dame University at the end of the war. Tom was stationed in Japan as the war ended and wrote about seeing damage caused by the atomic bomb at Nagasaki. He returned to the family car business after the war. Ann married a Wyoming rancher and business owner.

Harold Emerson Kokjer (November 18, 1898-July 25, 1981) was my grandfather. He went by Emerson or Em. Laura Winifred Clark Kokjer (July 16, 1900-April 25, 1966) was his wife. He was an Army radio sergeant in World war I and was stationed in Hawaii and France. He graduated with a law degree in 1922 and was offered a job in a law practice in Wahoo, Nebraska, about thirty miles north of Lincoln. Emerson and Winifred lived there the rest of their lives. He was town's mayor for one term. From 1939 to 1947, he commuted to Lincoln, where he was the deputy attorney general. He became a state district judge in 1947, and was later the official reporter for the Nebraska Supreme Court.

Winifred disliked her first name and went by Winnie or Win. After her children were grown, she was the society editor for the *Wahoo News*. She grew up in Lincoln and graduated from the University of Nebraska, where she and Emerson met.

They had two children:

Phyllis Marcia Kokjer Beck, was my mother (January 2, 1926 –November 9, 2019). Phyllis was six years younger than Ty. As soon as she was old enough, she worked at the ordnance plant in Mead, Nebraska, just east of Wahoo. She drove an electric truck and crawled into bomb casings to install fuses. She graduated from a two-year college in Wahoo, entered the University of Nebraska in 1945, and worked for an advertising agency after graduating in 1947. She and Leo J. Beck Jr. (October 6, 1923-January 12, 1997) married in 1948. They lived in Lincoln and had five children.

Janet Elizabeth Kokjer Sothan (June 2, 1931) was 11½ years younger than Ty. She graduated from the University of Nebraska and married Norman L. Sothan, a U.S. Navy aviator. She became an expert in genealogy, researching our family and teaching classes. They had three children. They moved often during Norm's career, then settled in Denver when he retired. After his death, Janet moved to Lincoln, where she still lives.

Ulysses S. Cobb (November 10, 1873–December 8, 1947) was Shy's father. He was a manager at a grain elevator in Funk, Nebraska, a railroad village. The population was 140 in 1940, and in 2021, it was 174. It's twenty-four miles southwest of Kearney and seven miles east of Holdrege. His family came to Nebraska from Indiana when he was a child. He wrote a number of letters to Ty, usually signed Gramps or Gramp. Augusta Huldeen Cobb (September 26, 1871–August 14, 1940) was Shy's mother. She was born in Sweden and immigrated to Nebraska with her family in 1880. She died just before Ty started pilot training. A couple of letters from her to Shy in 1939 and 1940 survive. The Cobbs had four children Robert, Shy, and two sons who died when they were young.

Robert Cobb and his wife, Grace, had three children: Robert (Bobby), Chandler (Chan), and Margaret. Robert was Shy's older brother. They lived in Alliance, Nebraska, west of Hyannis. The two families saw each other often after Hans and Shy moved to Hyannis. Ty was close to Chan and Bobby and thought of Margaret as the younger sister he never had. During the war, the family moved to Hastings, Nebraska, where Robert had a war production job, likely at the Navy Ammunition Depot. Bobby joined the Signal Corps, and Chan was in the Navy, on a supply ship in the Pacific. Margaret entered the University of Nebraska in September 1945.

Pete Huldeen/Holden was Shy's uncle on her mother's side. He lived in Honolulu, Hawaii, where he rented out holiday cottages and wrote occasional articles for local newspapers. Ty saw him when his ship stopped there on the way to Manila. After the war started, Hans and Shy urged Pete to return to the mainland, but he remained in Hawaii. Augusta's maiden name is listed as Huldeen, but Pete used Holden as his last name.

References

Allen, Oliver "Red." *Abandoned on Bataan* (2002) as told to Mildred Allen. Crimson Horse.

Ashcroft, Bruce. *We Wanted Wings: A History of the Aviation Cadet Program*, (Reprint edition 2009), staff historian for the Air Education and Training Command. Online: https://media.defense.gov/2015/Sep/11/2001329827/-1/-1/0/AFD-150911-028.pdf

Coleman, John H. *Bataan and Beyond: Memories of an American POW* (1978) Texas A&M University Press.

Dallek, Robert. *Franklin D. Roosevelt and American Foreign Policy, 1932-1945* (1995) Oxford University Press.

Daws, Gavin. *Prisoners of the Japanese: POWs of World War II in the Pacific* (1994) William Morrow and Company, Inc.

Dooly, Major Thomas. *To Bataan and Back: The World War II Diary of Major Thomas Dooly*, edited by Jerry C. Cooper (2016) Texas A&M University Press.

Dyess, Lt. Col. William E. and Leavelle, Charles. *The Dyess Story: The Eye-Witness Account of the Death March From Bataan, Japanese Prison Camps and Escape* (1944, 2002) G.P. Putnam's Sons).

Edmonds, Walter D., *They Fought with What they Had: The Story of the Army Air Forces in the Southwest Pacific, 1941-1942* (1951) Little Brown And Company. Online: https://media.defense.gov/2010/Oct/01/2001329751/-1/-1/0/AFD-101001-051.pdf

Eisner, Peter. *MacArthur's Spies: The Soldier, the Singer, and the Spymaster Who Defied the Japanese in World War II* (2017) Penguin Books.

Gause, Major Damon "Rocky." *The War Journal of Major Damon "Rocky" Gause: The Firsthand Account of One of the Greatest Escapes of World War II* (1999) Hyperion,

Goodwin, Doris Kearns. *No Ordinary Times, Franklin & Eleanor Roosevelt: The Home Front in World War II* (1994). Simon & Schuster.

Goralski, Robert. *World War II Almanac 1931-1945* (1981) G.P. Putnam's Sons, New York.

Groom, Winston. *1942: The Year That Tried Men's Souls* (2005) Atlantic Monthly Press. Kindle Edition.

Hardee, David L. *Bataan Survivor: A POW's Account of Japanese Captivity in World War II*

Edited by Frank A. Blazich, Jr. (2016) University of Missouri Press.

Heisinger, Duane. *Father Found: Life and Death as a Prisoner of the Japanese* (2003)

Xulon Press.

Hersey, John. *Men on Bataan* (1942, Fifth edition1943), Alfred A. Knopf.

Jacobs, Eugene C. *Blood Brothers*, (2012), Guttenberg e-Book, http://www.gutenberg.org/ebooks/8423 Also available in print, Alpha Editions, 2018.

Jones, Betty B. *The December Ship: A Story of Lt. Col. Arden R. Boellner's Capture in the Philippines, Imprisonment, and Death on a World War II Japanese Hellship* (1992) MacFarland & Company.

Knox, Donald. *Death March: The Survivors of Bataan* (1981) Harcourt Brace Jovanovich.

Lawton, Manny. *Some Survived: From a Soldier Who Was There … The Shocking True Story of the Bataan Death March* (1989) Warner Books.

Manchester, William. *American Caesar: Douglas MacArthur 1880-1964* (1978) Back Bay Books.

Michno, Gregory F., *Death on the Hellships: Prisoners a Sea in the Pacific War* (2001) Naval Institute Press.

Miller, Col. E.B. *Bataan Uncensored: The True Story of the Death March and the Subsequent Horrors of Japanese Prison Camps* (1949, 2018) Hart Publications, Inc.

Montgomery, Ben. *The Leper Spy: The Story of an Unlikely Hero of World War II* (2017) Chicago Review Press.

Morrill, John. *South From Corregidor: The Remarkable Story of a Terrifying Escape* (1943, 2018) Simon and Schuster. Kindle edition.

Morton, Louis. *The Fall of the Philippines*, (1953) Center of Military History, United States Army. The 50th anniversary edition is available in print. Online: https://history.army.mil/books/wwii/5-2/5-2_Contents.htm

Norman, Elizabeth M. *We Band of Angels: The Untold Story of American Nurses Trapped on Bataan by the Japanese,* (1999) Pocket Books.

Norman, Michael and Norman, Elizabeth M. *Tears in the Darkness: The Story of the Bataan Death March and Its Aftermath* (2009) Farrar, Strauss and Giroux.

Pearson, Judith L. *Belly of the Beast: A POW's Inspiring True Story of Faith, Courage, and Survival Aboard the Infamous WWII Japanese Hell Ship Oryoku Maru* (2001) New American Library.

Perry, Mark. *The Most Dangerous Man in America: The Making of Douglas MacArthur* (2014) Basic Books

Phillips, Claire. *Agent High Pockets* (1947, 20180) Binfords & Mort.

Reynolds, Robert V. *Of Rice and Men*, (1949) The Leicht Press.

Rovere, Richard H., and Schlesinger, Arthur, Jr., *The MacArthur Controversy And American Foreign Policy* (1951, 1965) Noonday Press, a division of Farrar, Straus and Giroux.

Schaefer, Chris. *Bataan Diary An American Family in World War II, 1941-1945* (2004) Riverview Publishing.

Schaller, Michael. *Douglas MacArthur: The Far Eastern General*, (1989) Oxford University Press.

Scott, James M. *Rampage: MacArthur, Yamashita, and the Battle of Manila*, (2018) W.W. Norton and Company.

Sides, Hampton. *Ghost Soldiers: The Epic Account of World War II's Greatest Rescue Mission*, (2001) Anchor Books.

Sloan, Bill. *Undefeated: America's Heroic Fight for Bataan and Corregidor* (2013) Simon and Schuster Paperbacks.

Stewart, Sidney. *Give Us This Day: The Powerful True Story of the Bataan Death March*, (1958) Popular Library.

Tenney, Lester. *My Hitch in Hell: The Bataan Death March*, (2000) Potomac Books, Inc.

Toland, John. *The Rising Sun: The Decline and Fall of the Japanese Empire*, (1970) Bantam Books.

Underwood, Charles Jr. *Deadline—Captain Charlie's Bataan Diary*, (2015) Piscataqua Press, a project of RiverRun Bookstore.

Wainwright, General Jonathan M. *General Wainwright's Story: The True Story of an American General Whose Courage Turned Brutal Defeat Into Final Victory*, (1945) Bantam Books

Weller, George. *Cruise of Death*, (1945) Chicago Daily News. Kindle edition (2015)

White, W.L. *They Were Expendable* (1942) Harcourt, Brace & Company. The book was made into a movie starring Robert Montgomery, John Wayne, and Donna Reed in 1951.

Whitcomb, Edgar G. *Escape from Corregidor*, (1958) Paperback Library: A Kinney Service Company.

Periodicals, papers, reports and unpublished works

Bjoring, Robert G., "An Autobiography," (undated). The self-published book is part of papers donated to the history office at Randolph Air Force Base after Bjoring's death.

Boothe, Clare, "MacArthur of the Far East: If War Should Come He Leads the Army That Will Fight Japan," *LIFE*, December 8, 1941.

Hansen, Maria and Osterhout, Elva Kokjer. "Kokjer Relatives," a family history. Circa 1967

Hobbs, Frank and Stoops, Nicole. Demographic Trends in the 20th Century: Census Special Report, 2002 https://www.census.gov/prod/2002pubs/censr-4.pdf

Kokjer, Hans Jr. and Charlotte "Shy," unpublished letters and scrapbook. In the collection of the Veterans History Project, Library of Congress.

https://memory.loc.gov/diglib/vhp/bib/ loc.natlib.afc2001001.117698

Kokjer, Madsen "Ty" Cobb, unpublished letters and diary. In the collection of the Veterans History Project, Library of Congress.

https://memory.loc.gov/diglib/vhp/bib/ loc.natlib.afc2001001.117698

File "Diary, Lt Madsen Cobb Kokjer, Folder 6," 17th Bombardment squadron, 27th Bombardment Group, letters home from Fukuoka POW Camp, April-November 1942, file code 999-2-89, Bk. 1.File "Diary, Lt Madsen Cobb Kokjer, Folder 6," 17th Bombardment squadron, 27th Bombardment Group, letters home from Fukuoka POW Camp, April-November 1942, file code 999-2-89, Bk. 1

https://catalog.archives.gov/search?q=Kokjer

[The title of Ty's diary erroneously states it was composed in Fukuoka POW Camp. The diary was written entirely while Ty was in the Philippines. He was transported to Japan, but was never at a Fukuoka POW Camp. As of this printing, National Archives and Records Administration in College Park Maryland, remains closed. Once it reopens, I hope to have this error corrected.]

File "Diary, Lt Madsen Cobb Kokjer, July-August 1942, Folder 7," letters home, file code 999-2-89, Bk. 2.File "Diary, Lt Madsen Cobb Kokjer, July-August 1942, Folder 7," letters home, file code 999-2-89, Bk. 2

https://catalog.archives.gov/id/12445095

"The Gunter Hotel In San Antonio's History," is an undated booklet published by the hotel. It has a brief chapter about the cadet club during World War II and a few photos.

Interviews

Darcy Pattison, daughter of Bataan veteran and Elleen Hutcheson. They are sisters and the daughters of a Bataan veteran.

K.C. Schaack, owner of the Hotel Hyannis until 2021.

Dan Crowley, Bataan Veteran

https://www.defense.gov/News/News-Stories/Article/Article/2463605/world-war-ii-vet-pow-who-endured-hell-ship-gets-cib-promotion-pow-medal/

https://www.legacy.com/us/obituaries/hartfordcourant/name/daniel-crowley-obituary?id=18463761

Joe Wilson, retired Army colonel and the son of Bataan veteran Colonel O.O. Wilson, who was on the Oryoku Maru. https://theeagle.com/news/local/the-life-of-ovid-o-wilson/article_4e860ec7-6a58-50bc-b8e7-435faa5ab7ed.html

Newspaper and magazine articles

Christenson, Sig. "POW's diary details Japanese atrocities: Doctor's diary told of Bataan," San Antonio News Express, May 24, 2015 Updated: May 24, 2015 9:11 p.m.

https://www.expressnews.com/news/local/article/POW-s-diary-details-Japanese-atrocities-6284566.php

A story about a POW who was on the Oryoku Maru, kept a diary, survived the way, stayed in the Army, and became a dental surgeon and professor. Dr. Bodine died May 17, 2005 at the age of 94 https://www.legacy.com/obituaries/sanantonio/obituary.aspx?n=roy-l-bodine&pid=88855737

Clark, Rob. "The life of Ovid O. Wilson: On Veterans Day, we remember a military hero with local roots," The Eagle, (Bryan, Texas) Nov 11, 2018.

https://theeagle.com/news/local/the-life-of-ovid-o-wilson/article_4e860ec7-6a58-50bc-b8e7-435faa5ab7ed.html

Zero Wilson, as he was known, and his family are also profiled in *Bataan Diary: An American Family in World War II, 1941-1945*, by Chris Schaefer.

End Notes

Introduction

1 Dallek, Robert. *Franklin D. Roosevelt and American Foreign Policy, 1932-1945.* Oxford University Press, 1995, 1979.

222 – U.S. didn't have ships to send to European allies. Roosevelt "pointed to the presence of the American Fleet in Hawaii as a deterrent to Japan."

236 – FDR told U.S. Ambassador to Japan Joseph Grew to tell Tokyo in the fall of 1939 that the U.S. would not be forced out of China, and would support China by reinforcing Manila and Pearl Harbor.

303 – FDR hoped expansion of air power in Philippines would deter Japan.

2 U.S. takes control of the Philippines in 1902, Library of Congress Research Guides, "Philippine-American War: Topics in Chronicling America." https://guides.loc.gov/chronicling-america-philippine-american-war

Morton, Louis. *The War in the Pacific: The Fall of the Philippines.* Center of Military History, United States Army, 1953.

88 – "The catastrophe of Pearl Harbor overshadowed at the time and still obscures the extent of the ignominious defeat inflicted on American air forces in the Philippines on the same day."

3 Goralski, Robert. *World War II Almanac : 1931-1945: A Political and Military Record.* G.P. Putnam's Sons, New York. 1981

403 – May 3, 1945, Davao is cleared.

413 – July 5, 1945, MacArthur declares victory in Japan.

4 File: "Diary, Lt. Madsen Cobb Kokjer, Folder 6," 17th Bombardment squadron, 27th Bombardment Group, letters home from Fukuoka POW Camp, April-November 1942, file code 999-2-89, Bk. 1.File "Diary, Lt Madsen Cobb Kokjer, Folder 6," 17th Bombardment squadron, 27th Bombardment Group, letters home from Fukuoka POW Camp, April-November 1942, file code 999-2-89, Bk. 1 https://catalog.archives.gov/id/12445059

[The title of Ty's diary erroneously states it was composed in Fukuoka POW Camp. The diary was written entirely while Ty was in the Philippines. He was transported to Japan, but was never at a Fukuoka POW Camp. The National Archives and Records Administration in College Park Maryland, was closed during the pandemic. I have asked to have the error corrected.]

File "Diary, Lt Madsen Cobb Kokjer, July-August 1942, Folder 7," letters home, file code 999-2-89, Bk. 2.File "Diary, Lt Madsen Cobb Kokjer, July-August 1942, Folder 7," letters home, file code 999-2-89, Bk. 2 https://catalog.archives.gov/id/12445095

5 Madsen Cobb Kokjer Collection, Veterans History Project, American Folklife Center, Library of Congress https://memory.loc.gov/diglib/vhp/bib/loc.natlib.afc2001001.117698

6 Norman, Michael L. and Norman, Elizabeth M. *Tears in the Darkness: The Story of the Bataan Death March and Its Aftermath.* Farrar, Straus and Giroux. 2009

38-41 – Ill-prepared troops, misperceptions.

Yam, Kimberly. "These Anti-Japanese Signs From World War II Are A Warning Against Bigotry Today," Huff Post. December 7, 2017

https://www.huffpost.com/entry/pearl-harbor-japanese-americans_n_5a283fb8e4b02d3bfc37b9f6

Davidson, Lucy. "5 Examples of Anti-Japanese Propaganda During World War Two," History Hit, Aug. 10, 2018.

https://www.historyhit.com/examples-of-anti-japanese-propaganda-during-world-war-two/

"The Era of Gangster Films," American Experience, PBS. Video, no author, undated.

https://www.pbs.org/wgbh/americanexperience/features/dillinger-era-gangster-films/

Rhodin, Erica and Glaser, Linda B., "1930s gangster films diversified conceptions of 'American-ness,' argues professor," Cornell Chronicle. October 26, 2011

https://news.cornell.edu/stories/2011/10/gangster-movies-offer-views-capitalism-exclusion

"Japan's 70-year struggle against Hollywood film stereotypes," Japan Today, Aug. 13, 2015

https://japantoday.com/category/features/kuchikomi/japans-70-year-struggle-against-hollywood-films-stereotypes

Morgan, Thad, "How Hollywood Cast White Actors in Caricatured Asian Roles:

Mickey Rooney's portrayal in 'Breakfast at Tiffany's' is often cited as offensive and a well-known example of yellowface," March 18, 2021. History Stories/History.com

https://www.history.com/news/yellowface-whitewashing-in-film-america

Ranger, Stephen, "Target Hollywood! Examining Japan's Film Import Ban in the 1930s," Wiley Online Library, July 8, 2020.

https://onlinelibrary.wiley.com/doi/full/10.1111/1758-5899.12818

Groom, Winston. *1942: The Year That Tried Men's Souls*, Grove/Atlantic, Kindle Edition, 2005

55 – 56 – Newsreels depicted Japanese people as cruel, yet Americans still liked Mr. Moto mysteries, which starred Peter Lorre, as a benign, and often helpful character.

Sloan, Bill. *Undefeated: America's Heroic Fight for Bataan and Corregidor*. Simon and Schuster Paperbacks, 2012

190-191 - Hatred of Americans was deeply ingrained in Japanese soldiers and became more intense after three months of fierce resistance by U.S. and Filipino soldiers. Japan hadn't encountered anything like it. Racism from propaganda convinced Japan that "Americans were crude, thuggish people bent on the destruction of the Japanese way of life."

Ruane Michael E. "Greatest Generation' runs counter to its wholesome image in survey on race, sex and combat during World War II," Washington Post, December 20, 2021.

https://www.washingtonpost.com/history/2021/12/20/greatest-generation-survey-race-sex/

The article describes a research project, "The American Soldier in World War II," directed by Edward J.K. Gitre, an assistant professor of history at Virginia Tech. The study found that many American soldiers returned from war wishing to preserve racism in the United States.

https://hci.icat.vt.edu/research/the-american-soldier-in-world-war-ii.html

Manchester, William. *American Caesar: Douglas MacArthur 1880-1964*, Back Bay Books, 1978

237 – Some Filipino soldiers wore helmets made from cocoanuts.

Martin Adrian R. and Stepheson, Larry W. *Operation Plum: The Ill-Fated 27th Bombardment Group and the Fight for the Western Pacific*. Texas A&M University Military History Series, 2008

35 – Soldiers were convinced by U.S. propaganda they could easily beat Japan.

Chapter One

7 Nature Conservancy, The. https://www.nature.org/en-us/about-us/where-we-work/united-states/oklahoma/stories-in-oklahoma/bison-history/ There were an estimated 30 million

bison, often called buffalos, when Columbus landed in America. They ranged from the Appalachian to the Rocky Mountains. By 1905, there were just over 1,000, about a quarter of them in captivity. By 2019, the number had risen to 350,000, most in private conservancies.

Kearney, Nebraska, https://www.cityofkearney.org/759/History

The city was named after Fort Kearny, which was named after Colonel Stephen Watts Kearny. He distinguished himself in the war of 1812 and the Mexican American War. The additional letter E is presumed to be a spelling error. http://uipress.lib.uiowa.edu/bdi/DetailsPage.aspx?id=202

8 Grant County population, 2020 Census, Nebraska Demographics by Cubit.

Grant County ranked 89th in population among Nebraska's 93 counties.

https://www.nebraska-demographics.com/counties_by_population

Grant County's population rose from 614 in 2010 to 722 in 2020. In 1940, the county's population was 1,327.

Grant County History

http://grant.mipsweb.info/webpages/about/history.html

Hyannis Nebraska, Data USA

https://datausa.io/profile/geo/hyannis-ne

9 Abbot Ranch https://www.abbott-ranch.com/ The ranch sells beef and welcomes visitors. A video about the ranch and the Sandhills includes a quick shot of the Hotel Hyannis at 1minute 52 seconds.

10 Postage rates in the 1930: letters 3 cents, postcards a penny.

https://about.usps.com/who-we-are/postal-history/domestic-letter-rates-since-1863.htm

"City School Notes," Kearney Weekly Tribune, April 18, 1929. Ty was one of 14 members of a fourth-grade team that had perfect scores in a multi-week spelling contest.

11 Torrey, Volta, "Hyannis, The Richest Town in America," Sunday World Herald, May 17, 1931. The hotel was owned by committee of ranchers.

http://genealogytrails.com/neb/grant/hyannisrichestwotninamerica.htm

12 Ellylson, Tyler, "Nebraska tuberculosis hospital documentary airing on NET," The Grand Island Independent, by UNK Communications, Feb 2, 2020 Updated Mar 11, 2020

The buildings were part of the Nebraska State Hospital for Tuberculosis in Kearney, 1912-1972. Now they are part of the University of Nebraska-Kearney. One of the buildings houses an exhibit about the hospital and the disease.

https://theindependent.com/life/entertainment/nebraska-tuberculosis-hospital-documentary-airing-on-net/article_3ec2d274-4644-11ea-ae7c-df2477bb9358.html

These three websites describe the State Industrial School, a home for "wayward children" that opened in Kearney in1881. "Industrial Building, State Industrial School, Kearney," Nebraska Memories, Nebraska Library Commission. Photo of school building.

http://memories.nebraska.gov/cdm/singleitem/collection/nlc/id/116/rec/3

"Group of buildings, State Industrial School, Kearney," Nebraska Library Commission.

http://memories.nebraska.gov/cdm/ref/collection/nlc/id/115

Doak, Susan, Southwest Nebraska Genealogy Society, "The Nebraska State Reform School,"

McCook Gazette, November 25, 2016. https://www.mccookgazette.com/story/2363327.html

13 Associated Press, "Reflects Farm Labor Shortage." Hastings Daily Tribune Feb. 23, 1942. Farmers are doing more work, also using tractors instead of horses, not baling hay but leaving it in piles on the ground, to get around shortage. Anyone who wants a farm job can get one.

Groom 54 – Farm hands, cooks and domestic servants leaving for defense jobs, which paid more. This threw off some segments of the economy. Still, four million Americans were out of work.

14 NCC Staff, "Five interesting facts about Prohibition's end in 1933," December 5, 2020, The National Constitution Center.

https://constitutioncenter.org/blog/five-interesting-facts-about-prohibitions cnd-in-1933

Hans Madsen Kokjer, 1857-1932 - [RG1141.AM]. Nebraska Historical Society.

https://history.nebraska.gov/collections/hans-madsen-kokjer-1857-1932-rg1141am

Ty's grandfather, Hans Madsen Kokjer, disapproved of drinking and smoking. Hans and Shy Kokjer weren't big fans of either, but they sold alcohol at the hotel to generate income.

15 University of Nebraska Cornhusker yearbook

1938 – 257, Sigma Chi. 302 – NU-MEDS

1939 – 134, Field Artillery Battery C. 248 – Sigma Chi

http://yearbooks.unl.edu/index.php?years=1920_1939

Cornhusker Marching Band History, Timeline. https://band-history.unl.edu/timeline

The band included women during the war, but reverted to all-male musicians until 1972, after female students threatened to complain to the faculty senate. The first female twirler was selected in 1961. She and her immediate successors were called "Sunshine Girls."

Daily Nebraskan, Nov.7, 1970, "Women band against band," https://nebnewspapers.unl.edu/lccn/sn96080312/1970-11-09/ed-1/seq-1/

Gainey, Sean A., retired brigadier general, "Making the best military officers in the world," U.S. Army History. ROTC was required at Land Grant universities until the 1960s, when it was made optional. https://www.cadetcommand.army.mil/history.aspx

16 Clarks Enterprise, Clarks, Nebraska. February 27, 1920

Obituary, Malina Alice Hartwell Kokjer. Tom Kokjer, her son, was recovering from injuries in San Antonio and missed his mother's funeral. https://images.findagrave.com/photos/2018/158/58220534_7a3254cc-5319-418d-b9f6-3ed935808691.jpeg

Dallek 172-173 – In a 1939 conference with military leaders, President Roosevelt asked for more ships and planes. His request for 10,000 planes was whittled down to 3,000. Half were for training pilots and the other half for potential combat.

221 – The advent of air power meant planes flying between 200 and 300 miles per hour could attack the United states, and vast oceans were no longer enough of a defense for the United States.

17 Goralski 2 - Japan attacks China's 7th Brigade at Mukden, September 18-19, 1931. Japan attacks rail line between Tientsin and Peking and occupies both cities. July 25-31, 1937.

64 - German troops entered Austria at dawn March 11, 1938.

90 – Germany invaded Poland Sept. 1, 1939.

Dallek - 193 – Japanese actions in China, and U.S. complaints.

198 – Germany invaded Poland.

Toland, John. *The Rising Sun: The Decline and Fall of the Japanese Empire*. Bantam Books, 1970.

507 – Japan's goal, the creation of The Greater East Asian Co-Prosperity Sphere, under Japan's leadership.

18 Dallek 215 – Roosevelt talked about peace and war.

19 Goralski 108-109 – Denmark and Norway fell after Germany invaded April 9, 1941, coincidentally, exactly one year before the fall of Bataan.

20 "A Heritage of Heroes—1940-1949" Jeep. https://www.jeep.com/history/1940s.html Perry, Susan, "WWII Scrap Metal Drive," Nebraska Prairie Museum, November 2007. This story focuses on Phelps County, where Holdrege and Funk are located.

https://nebraskaprairiemuseum.com/2020/04/13/wwii-scrap-metal-drive/

21 Justman, Ben, Omaha World Herald, Feb 21, 2016 Updated Jul 15, 2020

"Wanted for victory: your scrap metal" https://omaha.com/community/bellevue/wanted-for-victory-your-scrap-metal/article_24a7729f-fc6a-5354-b3d0-1f59508d1eb9.html

Duggan, Joe, Lincoln Journal Star, March 18, 2007, "Remembering the 1942 Nebraska Scrap Metal Drive" https://journalstar.com/news/local/remembering-the-1942-nebraska-scrap-metal-drive/article_44b311bb-ca27-556b-b4d4-716a0fd9b801.html

Rockoff, Hugh, Department of Economics, Rutgers University. "Keep on Scrapping: The Salvage Drives of World War II," National Bureau of Economic Research. Sept. 17, 2007

https://papers.ssrn.com/sol3/papers.cfm?abstract_id=1014795

Newspaper, Scrap Plan, Nebraska State Historical Society. The museum website features clippings from the Omaha World Herald, from August 1942, describing statewide scrap drives, that Grand County won, and that the paper won Pulitzer Prize in 1943 for public service.

https://nebraskahistory.pastperfectonline.com/webobject/A2256060-0887-420E-961D-291067347285

Omaha World Herald, 1943, Pulitzer Prize for Public Service.

https://www.pulitzer.org/prize-winners-by-year/1943

Chapter Two

22 Ashcroft, Bruce. *We Wanted Wings: A History of the Aviation Cadet Program*

HQ AETC/HO (Air Education and Training Command, Office of History and Research). Reprinted Edition 2009.

29-30 – expansion of the Army Air Corps.

23 Ashcroft 28 - Stearman PT-13 Kaydet, principal training plane.

Boeing-Stearman   https://www.boeing.com/history/products/stearman-kaydet-trainer.page

24 Ashcroft 25 – In recruiting, a majority failed initial tests and physical exams, which lowered the rate of crashes during training.

28 – expected graduation rate was 50 percent

25 Metzger, Marvin Irving. Obituary, Lincoln Journal-Star, January 7, 2011

https://journalstar.com/lifestyles/announcements/obituaries/metzger-marvin-irving/article_e3816b03-45ac-5dab-bd15-0b7ac366a37a.html

He saw action at Casablanca, Guadalcanal, New Guinea, Iwo Jima, and Okinawa.

NBC Radio City West https://eyesofageneration.com/studios-page/nbc-studios-los-angeles/

NBC opened its West Coast Studios in 1927 https://populartimelines.com/timeline/NBC

Radio City headquarters opened in Los Angeles in 1933 http://www.radiocityhollywood.com/

26 Groom 28-29 – Poll: 70 percent of Americans did not want war. Selective service act 1940, draft of 900,00 men for one year without added approval by Congress. Neutrality Act, 1936, could not aid or trade with warring powers.

Toland 3 – population explosion in Japan.

5-10 – Japan's takeover of Manchuria.

72-73 – Tripartite Pact.

Goralski – 2-3 – September 19 to Oct. 9, 1931, Japan takes over Manchuria.

116 – Dunkirk evacuation, May 26 to June 4, 1940.

118 – Italy entered the war.

122 – France surrendered to Germany, June 21, 1940.

132 - Sept. 27, 1940, Germany, Italy, Japan sign Tripartite Pact.

Dallek 105 – The United States favors neutrality

108 – Neutrality Act.

27 Sparrow, Paul M., Director, FDR Library. "The Most Important Presidential Election in History." November 2, 2015.

https://fdr.blogs.archives.gov/2015/11/02/the-most-important-presidential-election-in-history/

28 Ashcroft 34 – weight limits for pilots.

Chapter Three

29 Randolph Field Historic District: Aviation: From Sand Dunes to Sonic Booms. National Park Service. https://www.nps.gov/articles/randolph-field-historic-district.htm

Randolph "played an exceptional role in the development of the air arm of the U.S. Army," according to the National Park Service, which maintains the Randolph Historic District that is now part of Joint Base San Antonio and trains aviation instructors.

Randolph Air Force Base Training (1941)

https://www.youtube.com/watch?v=l1louwV-PY0

30 "Scene at Randolph, Texas," Kearney Daily News, Feb. 11, 1941

DeWolf one of 10 from Nebraska to graduate from Randolph. Lists Ty as former Kearney resident. First at basic school, 65 hours of flying, half solo in "rugged primary training planes." At Randolph, they flew "racy low-wing monoplanes. 70 more hours of flying and learning all the maneuvers that Ty listed.

To Kelly next. Paid $75/month at Randolph, plus room, board, uniforms. When commissioned $205/month.

31 Ashcroft 41 – BT-13 and BT-15.

National Museum of the United States Air Force, North American BT-14 (NA-64). Museum photograph and description.

https://www.nationalmuseum.af.mil/Visit/Museum-Exhibits/Fact-Sheets/Display/Article/196918/north-american-bt-14-na-64/

32 The Gunter Hotel in San Antonio's History, booklet, undated, not paginated.

Published and distributed by the hotel, now the Sheraton Gunter Hotel.

Photos of the Cadet Club and a tea dance. "The Pan-American Room of the Cadet Club, and the veranda were reserved for the cadets. Invitations were sent out to select young ladies" for dances. "All during the war, Tea Dances at the Gunter were part of the social scene in San Antonio."

33 Air Force Civil Engineer Center, Former Kelly AFB, Texas https://www.afcec.af.mil/Home/BRAC/Kelly.aspx

Kelly Field Heritage http://kellyheritage.org/wwii-era.asp

In fact, by mid-war, students would graduate from the primary course to the advanced course in recognition that they received little training in what was initially the third phase, going directly from the PT-13/17/19/22s to the AT-6 [Boyd]

34 BC-1. All-Aero, North American NA-26 / BC-1 / T-6 / SNJ / Harvard / A-27 / SNJ. 2000 http://all-aero.com/index.php/53-planes-l-m-n-o/7364-north-american-na-26-bc-1-t-6-harvard

BC-1/AT-6 Condor Squadron, Officer's and Airmen's Association, Inc., The Legendary AT-6 "Texan" https://www.condorsquadron.org/at6/

Military Factory, North American T-6 Texan: Two-Seat Advanced Trainer Aircraft, photographs and description. https://www.militaryfactory.com/aircraft/detail.asp?aircraft_id=408

Link Trainer. Delta Flight Museum, Link Trainer Model AN-T-18 Flight Simulator. Photographs and description.

https://deltamuseum.catalogaccess.com/objects/2948

Bjoring, Robert G., unpublished memoir.

40 - He reported that cadets had to spend 12 hours in the Link Trainer, which turned on a pedestal and tilted up and down. It could simulate emergencies, and someone tracked their responses.

35 Smithsonian Air and Space Museum and Smithsonian Museum of American History, "Time and Navigation: The untold story of getting from here to there. Flying the Beam."

https://timeandnavigation.si.edu/multimedia-asset/flying-the-beam

36 Goralski 145 – Ambassador Grew reports rumor of possible attack on Pearl Harbor.

Toland 151-152 – State Department ignores Ambassador Grew's concerns.

173 – Grew hears rumors in January that Japan is thinking of attacking Pearl Harbor. The message was routed to Naval Intelligence, which discounted it. Meanwhile, Japan began to study such an attack.

37 Toland 170 – A war that need not be fought.

38 Goralski 146 – Germany begins to transport Polish Jews to the Warsaw ghetto.

Goralski 150 – Japan takes control of rice in Indochina. Final approval of U.S. Lend Lease Act March 11, 1941.

152 – Japan controls all rubber from Dutch East Indies.

153 - A more conservative cabinet takes office in Tokyo, April 10, 1941.

Toland 135 – General Hideki Tojo becomes prime minister, on the emperor's request, Oct. 16, 1941.

39 Goralski 156 – The U.S. sends 2,000 soldiers to the Philippines.

Calamur, Krishnadev, "A Short History of 'America First': The phrase used by President Trump has been linked to anti-Semitism during World War II."

The Atlantic, January 21, 2017

Perry, Mark. *The Most Dangerous Man in America: The Making of Douglas MacArthur*. Basic Books. 2014.

267 – Charles Lindbergh and P-38, unauthorized combat mission.

Eyewitness to History, "Charles Lindbergh in Combat, 1944"

As a consultant to the Fort Motor Company, Lindbergh, a civilian, tested the company's planes, and flew with military pilots on missions to observe how the planes were used.

http://www.eyewitnesstohistory.com/lindbergh2.htm

Whitman, Alden, "Lindbergh Says U. S. 'Lost' World War II," book review,

New York Times, Aug. 30, 1970

https://www.nytimes.com/1970/08/30/archives/lindbergh-says-us-lost-world-war-ii-lindbergh-contending-that-he.html

40 Mendoza, Madalyn, "America's last remaining Pig Stand celebrates 100 years in San Antonio this fall," MySA, July 29, 2021Updated: Aug. 2, 2021 3:19 p.m.

https://www.mysanantonio.com/food/restaurants/article/pig-stand-san-antonio-100-years-16347027.php

Chapter Four

41 Air Force History, Military.Com, 2021 https://www.afhistory.af.mil/FAQs/Fact-Sheets/Article/458985/evolution-of-the-department-of-the-air-force/

https://www.military.com/air-force-birthday/air-force-history.html

The U.S. military purchased its first aircraft in 1909. At the start of World War I, the United States had six planes and 14, pilots. One duty was to patrol the Mexican border in 1916.

The air service had several names and was split among several agencies, including the Signal Corps, before it became the U.S. Army Air Corps in 1926. Bombers and pursuit planes were stationed at the Panama Canal, Hawaii, and the Philippines. The total force in 1926 included fewer than 1,000 planes, 919 officers and 8,725 enlisted men. By 1938, there were 7,000 to 10,000 planes. By June 1942, there were 26,500 men in the Air Corps and 2,200 planes. And in 1945, there were 2,250,000 men and women and 63,715 planes. The service was renamed the Army Air Forces in 1941, and became the U.S. Air Force July 26, 1947.

"Dayton, Aviation, and the First World War," National Park Service

https://www.nps.gov/articles/dayton-aviation-and-the-first-world-war.htm

As World War I began in Europe, the United States had six planes and 14 pilots. Three years later, there were 200 planes, mostly trainers. But by 1919, that number rose to 4,500.

Glenshaw, Paul and Patterson, Dan, "America's First Combat Pilots: Descendants of the volunteers who served in the famed Lafayette Escadrille shed light on why they chose to fight."

Air & Space Magazine, December 2014 https://www.air-spacemag.com/military-aviation/americas-first-combat-pilots-180953371/

The first U.S. pilots to fly in war did so in French planes.

"General Henry H. Arnold," U.S. Air Force https://www.af.mil/About-Us/Biographies/Display/Article/107811/general-henry-h-arnold/

Henry A. "Hap" Arnold biography

42 Groom 53 – Americans "blissfully unaware" about likelihood of war, despite knowing the country was ramping up manufacturing. And despite increasing tension with Japan.

43 Schaller, Michael. *Douglas MacArthur: Far Eastern General,* 1989, Oxford University Press, 1989

45-46 – Japan's decision July 2 to move south.

Toland 86-87 – As of May 1941, the United States had broken Japan's diplomatic code.

44 Schaller 49 – MacArthur's plan to defend the entire coastline.

45 Toland 98 – Roosevelt freezes Japanese assets in the United States, cuts oil shipments, and returns MacArthur to active duty.

Dallek 273-275 – Roosevelt was under pressure the spring of 1941 to embargo oil shipments to Japan to relieve domestic shortages, but he decided against it to avoid causing Japan to attack Russia. No one knew if Japan would attack Russia on the west to help Germany, which had attacked on the east or do something else. Roosevelt: "Every little episode in the Pacific means fewer ships in the Atlantic." Instead, his actions provoked Japan. This and a harsher U.S. response to a previous Japanese diplomatic offer on June 21, "led to a high-level decision on July 2 to proceed with the advance to the south, even if it meant war with Britain and the United States."

Manchester 29-30 – Major General Arthur MacArthur in the Philippines.

64 – Douglas MacArthur's first assignment to the Philippines

127 – Douglas assigned to second tour of duty in the Philippines

167-168 – Plans for PT-Boat fleet.

Scott, James M., *Rampage: MacArthur, Yamashita, and the Battle of Manila,* W.W. Norton & Company, 2018.

25 – MacArthur spent part of his childhood in the Philippines, and it was his first assignment after West Point. He befriended Manual Quezon, Sergio Osmena, both at Santo Tomas, later the country's first two presidents.

46 Toland 144 – Japan set a deadline of Nov. 30, 1941, to end negotiations with the United States.

193-194 – Decision Nov. 5, formal preparations. Nov. 6. Japan knew the U.S. fleet was in the harbor on weekends and there would be a full moon.

Schaller 52 – The United States rejected Japan's latest offer Oct. 16, 1941. As a result, Prince Fumimaro Konoye resigned from the cabinet and General Hideki Tojo became prime minister.

53 – U.S. rejects Japan's final offer Nov. 26, acknowledging it likely meant war. But Japan had already decided—on Nov. 5—to attack in early December if the U.S. and Japan could not reach an agreement.

Groom 48 – winter weather meant window to attack Pearl Harbor was nearly over.

38-39 – code breaking.

Dallek 303-307 – Roosevelt tried to extend time for negotiations. Sea and air strength should have been a deterrent to Japan by mid-December. By early spring 1942, it could be the deciding factor. Japan continued talks with U.S. but at home, the deadline was the end of November.

Hersey, John. *Men on Bataan*. Alfred A. Knopf, 1942.

287-289 – portrait of MacArthur and his evaluation of Japan as an enemy.

Goralski 149 – Hitler urged Japan to be active in a Far East War, March 5, 1941.

Schaller 48 – MacArthur told Washington at the end of August 1941 he would be ready by April 1942, the earliest Japan could attack.

47 Wainwright, Jonathan M. *General Wainwright's Story: The True story of an American General Whose Courage Turned Defeat into Final Victory*. Bantam Books, 1945.

10-11 – The sparkle went out of Manila.

14-15 – Dependents evacuated, how ill-prepared the military was.

Morton 26 – shortage of equipment and personnel when war came.

Miller. E.B., *Bataan Uncensored: The True Story of the Death March and the Subsequent Horrors of the Japanese Prison Camps*, The Hart Publications Inc., 1949, 2018.

83 – praise for Philippine Scouts ("No finer soldier was ever found."), but the rest of the army was ill-prepared and had little training, few weapons. Many dialects meant communications were poor. "A motley force at best ... not to any fault of theirs."

48 Bjoring 432 – finds out he is going to Manila, Sept. 25, 1941.

49 Norman and Norman 9-10 – Gallup Poll published in October 1940.

Goralski 112, 154, 159, 167 – After the fall of Denmark in April 1940, Iceland declared independence from Denmark. In July 1941, U.S. Marines replaced British forces to prevent Germany from establishing air or naval bases there. The U.S. also assumed responsibility for the defense of Greenland, another Danish outpost.

Dallek 218-219 – "Germany's occupation of these Danish possessions would present a direct threat to Britain and all of North America." Roosevelt preferred U.S. occupation to that of Britain or Canada, fearing it would cause Japan to invade the Dutch East Indies or cause Germany to take over Holland.

267 – By May 1941, 85 percent of Americans assumed we would eventually enter the war, although 75 percent wanted to avoid war.

302 – Japan's demands by October required that the U.S. give up defending China and to help Japan find resources. In return, Japan would guarantee neutrality for the Philippines. Sixty-seven percent of Americans were willing to risk war with Japan to keep it from becoming too powerful.

50 National Park Service, Golden Gate Recreation Area. The San Francisco Port of Embarkation: Servicing the Army's Needs in the Pacific, April 8, 2019

https://www.nps.gov/goga/learn/historyculture/port-of-embarkation.htm

51 B-17 Flying Fortress: Historical Snapshot.

https://www.boeing.com/history/products/b-17-flying-fortress.page

B-25 Mitchell Bomber: Historical Snapshot

https://www.boeing.com/history/products/b-25-mitch-ell.page

National World War II Museum, *Boeing B-17E Flying Fortress*

https://www.nationalww2museum.org/visit/museum-campus/us-freedom pavilion/warbirds/b-17e-flying-fortress

52 Martin and Stephenson 30 – Food like Henry VIII would have eaten. "Little did those feasting like royalty know that, in a few short months, they would be starving on Bataan and considering themselves fortunate whenever they were able to obtain a morsel of meat from a monkey, water buffalo, or army mule."

53 Hardee, David L. *Bataan Survivor: A POW's Account of Japanese Captivity in World War II,* Edited by Frank A. Blazich, Jr., University of Missouri Press, 2016

11 – senior officers enjoyed the trip but knew they were going to war.

Dyess, Lt. Col. William E. *The Dyess Story: The Eye-Witness Account of the Death March From Bataan, Japanese Prison Camps and Escape*, Edited by Charles Leavelle, G.P. Putnam Sons, 1944

17 – Knew war was likely, despite most Americans laughing at the prospect.

Chapter Five

54 Eisner, Peter. *MacArthur's Spies: The Soldier, the Singer, and the Spymaster who Defied the Japanese in World War II.* Penguin Books. 2017

10 – Manila, elegant tropical capital, population of 600,000.

Sloan 3 – Manila, Pearl of the Orient

Pearson, Judith L. *Belly of the Beast: A POW's Inspiring True Story of Faith, Courage, and Survival Aboard the Infamous WWII Japanese Hell Ship Oryoku Maru.* New American Library, 2001.

29-30 – Easy life in Manila.

55 Schaefer, Chris, *Bataan Diary: An American Family in World War II, 1941-1945*, Riverview Publishing, 2004.

6-7 – smells at Pier 7.

Bjoring – 49 – smells at pier.

Aquino International Airport occupies the former Nielson Field site, then moved to the nearby Nichols Field site.

https://placeandsee.com/wiki/ninoy-aquino-international-airport

56 Sloan 1-3 – low salaries supported a luxurious life. Ralph Hibbs, life is easy here.

Norman, Elizabeth M. We *Band of Angels: The Untold Story of American Nurses Trapped on Bataan by the Japanese.* Pocket Books, (1999)

3-4 – Military nurses assigned to the Philippines in 1941 enjoyed a pleasant life. They could not imagine Japan would attack. "It was a halcyon life, cocktails and bridge at sunset, white jackets and long gowns at dinner, good gin and Gershwin under the stars."

57 Office of the Historian, U.S. Department of State, "The Philippine-American War, 1899–1902" https://history.state.gov/milestones/1899-1913/war

"The Philippines, 1898–1946," History, Art and Archives, The United States House of Representatives https://history.house.gov/Exhibitions-and-Publications/

APA/Historical-Essays/Exclusion-and-Empire/The-Philip-
pines/

History of US-PI. Main exports.

Schaefer 6-7 Smell of the port. Description of Pier 7.

Scott 427-428 – Post-war analysis of property damage
in Manila totaled $800 million. Lists pre-war conveniences,
homes, businesses, and treasures destroyed.

Montgomery, Ben, *The Leper Spy: The Story of an Unlikely
Hero of World War II*, Chicago Review Press, 2017.

5-7 – Description of Manila.

Schaller 23 – U.S imported cordage, coconut oil, sugar, all
without tariffs.

Sloan 3-5 – Description of Manila, including the poor neighborhoods.

Toland 144 – Japan set deadline for negotiations, Nov. 30,
1941.

216 – the fleet was on its way, headed north first to avoid
detection.

199 – Stimson message to MacArthur

Groom 51-52 – fleet on the way, spotted, but US assumed
it was going to Indochina

58 Hersey 16 – Time Magazine reported increasingly dire
news about possible attack. Some of Hersey's writing was
based on reporting done by colleagues, including Melville Jacoby and his wife, Annalee. They were among the reporters
who were evacuated to Australia. Melville died there in a freak
accident in April 1942. She later married writer and editor
Clifton Fadiman.

59 Schaller 54 – Boothe (Luce) interview with MacArthur
took place in October.

U.S. Army Center of Military History

Douglas MacArthur, list of postings and promotions. Promoted to general, or fourth star December 1941. https://his-
tory.army.mil/html/faq/macarthur.html

60 Sloan 17 – mail service between Manila and the United States about to be suspended.

The Pan Am Historical Association. How Pan Am Went to War.

https://www.panam.org/war-years/609-clippers-at-war

https://www.panam.org/war-years/400-how-america-s-airline-went-to-war-2

Norman and Norman 18 – incompetent Japanese soldiers, expecting a short war.

61 Toland 214 – figuring out how to get a message to the emperor.

62 Morton 71-72 – party to honor Brereton, entertainment like Minsky's. Dec. 6, nothing ominous, a normal day. Party like Minsky's

Edmonds, Walter D., *They Fought with What they Had*, The Center for Air Force History, Little Brown and Company, 1951

72 – Few noticed that Brereton slipped out for calls.

63 Toland 257 – Japan's ambassador calls on Secretary of State Hull.

64 Price, S.L. The Second World War Kicks Off December 7, 1941, Redskins Versus Eagles on Pearl Harbor Day, Sports Illustrated, November 29, 1999.

https://vault.si.com/vault/1999/11/29/the-second-world-war-kicks-off-december-7-1941-redskins-versus-eagles-on-pearl-harbor-day

Story of December, 7, 1941, football game and how fans learned of attack.

Toland 258-259 - radio listeners learned of attack. A few people chopped down cherry blossom trees that had been a gift to the U.S. capital from Japan. The Japanese embassy was surrounded by crowds. By evening the diplomats were relocated to a hotel under guard.

Heisinger, Duane, *Father Found: Life and Death as a Prisoner of the Japanese in World War II*, Xulon Press, 2003

109 – How his family learned his father was now at war.

Groom 97-98 – Dec. 8, recruiting, long lines. Documents —Constitution Declaration of Independence, Bill of Rights, Magna Carta, Gutenberg Bible—ordered to Fort Knox by the librarian of Congress. Sent by rail with armed guards.

87 – Cherry trees chopped down, crowds gathered.

Blakemore, Erin, What Happened to America's Most Precious Documents After Pearl Harbor? Smithsonian Magazine, December 6, 2016 https://www.smithsonianmag.com/smart-news/what-happened-americas-most-precious-documents-after-pearl-harbor-180961325/

"History of the Cherry Trees," Cherry Blossom Festival, National Park Service, February 13, 2021. https://www.nps.gov/subjects/cherryblossom/history-of-the-cherry-trees.htm

Wainwright 16 – Tension the night before the war started, and last decent sleep.

Schaller 55 – word about Pearl Harbor reached Manila.

Martin and Stephenson 52 – After hearing about Pearl Harbor: "Most of the pilots were sleeping off the effects of the wild party the night before." All assumed "an immense convoy" was on the way with planes and other equipment.

65 Alex, Dan, March 12, 2019, MilitaryFactory.com

Douglas A-24 Banshee: Dive Bomber / Reconnaissance Aircraft https://www.militaryfactory.com/aircraft/detail.asp?aircraft_id=491

Boeing, SBD-A-24 Dauntless Dive Bomber https://www.boeing.com/history/products/sbd-a-24-dauntless.page

Martin and Stephenson 12 – plane meant to fly toward target, firing machine gun, then drop a relatively large bomb.

18 – Dauntless converted from Navy to Army plane.

25 - They were promised the A-24s would be in Manila when they got there.

Manchester 203 – as of Dec. 7 in Manila, the *Pensacola* convoy was 4,887 miles away, near Kiribati, south of Hawaii.

212 – After the attack on Manila, the *Pensacola* fleet carrying planes for the 27th turned around and headed to Honolulu.

Morton 48 - In the convoy were the 52 dive bombers of the 27th Bombardment Group, planes for other groups, vehicles, guns, ammunition, bombs, aviation fuel, and soldiers.

146 – While the safety of the Hawaiian Islands was undoubtedly of prime importance, the decision to bring back the *Pensacola* convoy was, in effect, an abandonment of the Philippine Islands.

Bjoring 51 – disappointment of pilots at plane they were assigned to fly.

66 Edmonds 51 – never send troops to war without their equipment. The planes were initially useless when they got to Australia because of missing parts.

68 – available planes were all old.

67 Dooley, Major Thomas. *To Bataan and Back: The World War II Diary of Major Thomas Dooley*. Edited by Jerry C. Cooper. Texas A&M University Press. 2016

4-6 – Saw a radiogram the morning of Dec. 8 saying a state of war existed.

Morton 77-78 – Fall of Wake, Guam and other Pacific islands.

Edmonds 87-90 – debate about bombing Formosa attack.

68 Edmonds 81 – 5 a.m. request turned down

93 – the whole discussion is academic

69 Sloan 53 – MacArthur, without a doubt, the one person who could control the situation

Rovere, Richard H. and Schlesinger, Arthur Jr. *The MacArthur Controversy And American Foreign Policy*. The Noonday Press. 1951.

54 – MacArthur fixed on his plan to defend the entire country.

Manchester 206 – MacArthur the key to the riddle, but we don't know his thoughts.

70 Sloan 54 – MacArthur had authority to attack Japanese bases within range.

Perry

73-74 – By 11:30 planes on the ground, wingtip-to-wingtip.

71 Hersey 21 – describes bombs dropping on Clark Field.

72 Dooley 1 – thought fires were locals burning brush.

Eisner – 15 – war came slowly at first to Manila.

Sloan 52 – Quoting Tech Sergeant Robert B. Heer, a carpenter in the 19th Bomb Group: "We laughed at the Japanese pilots and planes. ...Then the Jap Zeros shot down our P-40s like chunks of firewood, and they caught us flatfooted at Clark because we'd never been taught want to do in case of an actual air raid."

73 Perry 86 – Japan landed a few troops Dec. 10

93 – 43,000 Japanese troops by Dec. 22.

Morton 88 – "The catastrophe of Pearl Harbor overshadowed at the time and still obscures the extent of the ignominious defeat inflicted on American air forces in the Philippines on the same day." Hopes for the active defense of the Islands rested on these aircraft that were destroyed and still om ships. At the end of the first day of war, such hopes were dead.

Sloan 43-44 – attack on Cavite, most of the fleet had left.

60 – 43,000 Japanese soldiers had joined the advance troops.

Norman and Norman 45 – main body of Japanese troops landed, describes Lingayen.

Schaller 56 – main Japanese assault Dec. 22, Lingayen Gulf.

Edmonds 132-133 – 27th could have had a field day.

74 Bjoring 53 – pilots work as plane spotters.

Edmonds 172-173 – what non-flying pilots and others in the 27th did in the first days of the war.

Martin and Stephenson 63 – Men at Fort McKinley assumed it would be next. Enlisted men dug foxholes, manned machine guns, watched the skies. Attack came at 3 a.m. Dec. 9.

75 Sloan 50 – stateside stories of victories fabrications or exaggerations.

76 Dyess 6 – He flew a few bombing missions

77 Norman and Norman 41-42 – size of country, untrained troops, unrealistic defense plans.

Martin and Stephenson 155 – PT boat plan and number of boats.

Dooley 13 – what U.S. Air Forces were, Japan will rue the day.

17 – to Manila, dancing at Manila Hotel.

25 – promoted to captain on New Year's Eve, later promoted to major.

78 Manchester 215 – Capital and general lived in a world of fantasy. MacArthur got his Fourth star Dec. 22

Perry 54 – WPO-3 was sheer nonsense, rescue would take two to three years.

79 Morton 58-59 – Japanese forces were confused when U.S. troops were in Bataan, not Manila.

Sloan 78 – Crucial Japanese mistake to bypass Bataan.

52 – Philippines, other Pacific islands declared expendable.

80 Hersey 77-79 – description of Bataan.

81 Edmonds 206 – Men evacuated as best they could on wildest day

Hardee 19-20 – description of fires, explosions as they evacuated from Manila to Bataan.

Martin and Stephenson 83 – "The pier itself was a solid mass of humanity and the lack of information was demoralizing." So many trucks, they couldn't leave once they were unloaded. Lots of boats on the bay in the dark, most running without lights. Boats were overloaded and there were no life jackets. They didn't know where the minefields were.

82 Groom 137-138 – food left behind in haste, soldiers would remember with bitterness.

Toland 289 – Bataan was bedlam.

Sloan 58 – Transferring enough food would have taken two weeks.

Manchester 215 – MacArthur could have moved 50 million bushels of rice

Norman and Norman 114-115 – Abandoned rice could have feed soldiers for years. MacArthur "had forgotten logistics." Number of people on Bataan.

83 Perry 122-123 – MacArthur's only visit to Bataan Jan. 10.

130-131 – Soldiers felt abandoned, they called MacArthur, whom most never saw, Dugout Doug.

Schaller 58 – Visit undermined morale, Dugout Doug.

Sloan 97 – briefly boosted morale, gave way to Dugout Doug.

Morton 375 – Most soldiers on Bataan believed those on Corregidor had more and better food. Still on half rations, but the diet was better balanced.

Knox, Donald. *Death March: The Survivors of Bataan.* Harvest/HBJ Book. 1981

79 – A shame that pilots were fighting as infantry.

84 Coleman, John S., Jr. *Bataan and Beyond: Memories of an American POW.* Texas A&M University Press, 1978

19-20 – His unit made several runs to Manila for food, ending on New Year's Eve.

85 Gause, Damon "Rocky." *The War Journal of Major Damon "Rocky" Gause: The Firsthand Account of One of the Greatest Escapes of World War II.* Hyperion, 1999

Gause 1-2 – New Year's Eve Party at Manila Hotel.

5-7 – escape from Manila.

Gause was a pilot in the 27th but there is no indication he and Ty knew each other. Gause did not surrender. He got fake papers identifying him as a Spanish-Filipino, found a boat, and sailed to Australia. He kept a ship's log and diary. He returned to the United States for about a year, staying for the birth of his son, then returned to duty in Europe. He was killed March 9, 1944, when his plane crashed as he practiced a new flight maneuver over the Isle of Wight. His Son, Damon L. Gause, found the log and published it.

Morton 188 – In just over a week, from Christmas Eve to New Year's day, Japan had gone from landing troops on northern Luzon to threatening Manila and Bataan.

Martin an Stephenson 88 – The week after Christmas the squadrons sent trucks back to Manila on a daily basis to pick up food and supplies." Manila was an open city but Japanese soldiers hadn't arrived. Allowed the 27th to bring in as much canned goods and rice as possible.

95 - After Jan. 1, no more supply runs to Manila. The Army blew up the Calumpit Bridge over the Pampanga River at 6:15 a.m. "The tons of supplies left inf Manila and Cabanatuan fell into enemy hands."

Chapter Six

86 Martin and Stephenson 88 – 17th Squadron camp at Limay.

The Radio Historian, edited by Jim R. Bowman, Manager KGEI, 1963-1977. International Broadcast Station KGEI: 1939-1994. History courtesy of FEBC International

http://www.theradiohistorian.org/kgei.htm

87 Morton 248 – provisional infantry, including air corps troops.

Martin and Stephenson 96 – Bomb group and others without planes were ordered to start infantry training. The pilots, mechanics, and technicians never had infantry training. They just got rifles.

108 – About airman fighting as infantry: SSgt. Jesse Knowles, "We knew about as much about that job as a goat knows about riding a bicycle."

Hardee 27 – assigned as executive officer of the Provisional Air Corps Regiment. It is possible that he was the lieutenant colonel Ty referred to as drilling them in infantry basics.

Hardee said it was a pity the pilots and others couldn't perform skills they had trained for.

Knox 100 – bomb group members "remotely acquainted" with weapons.

88 Morton 254-257 – Late decision to retreat, not enough planning to send supplies. A food inventory conducted Jan. 3, not enough calories. They had a 30-day supply for 100,000 men, including a 50-day supply of canned meat and fish, 40 days of canned milk, 30 days of flour and canned vegetables, plus some sugar, fruit, coffee, potatoes, and cereals.

368 – quarter rations by February.

89 Wainwright 45 – forage ran out, and they had to shoot and eat horses and mules.

Dooley 68 69 – shortage of food. He hasn't had to eat any of the horses or mules yet.

Morton 399-370 – quoting an officer about relative acceptability of various animals they ate. Rice became the main food, men thought it tasted like wallpaper paste. Mostly starch, it has few vitamins.

404 – combat efficiency rapidly approaching zero.

Dyess 36 – life expectancy of animals on Bataan was nil.

Montgomery 55 – Rations down to three-eighths. Men wasted away. Two open-air hospitals cared for 10,000 men.

Groom 147 – weight loss of soldiers. Both sides were exhausted by late January. American forces ran out of grain for horses, killed and ate them. "The food shortage, which had been critical was now acute." Soldiers who weighed 170-200 pounds down to 150 within a month. "Soon everyone's clothes hung off them like a scarecrow, that is, what was not already in tatters or rotten from the harsh jungle fighting." Also malaria, dengue fever, scurvy, beriberi, hookworm, amebic dysentery, edema, night blindness.

90 Morton 258 – shortages of shelters, mosquito netting, quinine.

91 Morton 222-225 – moving south, blowing up bridges behind them

230 – during the retreat, half of Filipino soldiers left and returned home.

Dooley 27 – Always falling back.

Martian and Stephenson 99 – The 27th moved to the Orion-Bagac reserve line

92 Dooley 40-41 – losing confidence. They are sacrifices to the great effort. Will report if he makes it through the war.

Grace, Tonya S., "Diaries tell forgotten chapter of WWII," The Kentucky New Era, Dec. 6, 2016. Feature about the notebooks and the book. Dooley retied from military in 1969 as a colonel and died in 2006. https://www.kentuckynewera.com/web/article_27508c74-bc47-11e6-98fd-b7e35a13a7ab.html

93 Toland 296 – Japan lowered U.S. flag and raised Japanese flag. Homma was convinced he had won the battle, that troops on Bataan were disorganized remnants of the U.S. Army.

Goralski 196 – Japan occupies Manila.

Groom 143 – Jan. 2 , Japanese troops arrived in Manila. Hauled down U.S. flag, Japanese soldier stomped on it, raised Rising Sun. Homma moved into MacArthur's suite at the Manila Hotel. Hung flag from balcony "General Homma moved himself into MacArthur's sumptuous penthouse apartment in the Manila Hotel and installed on its balcony an enormous Rising Sun flag, which could be seen, with great aggravation, with field glasses on clear days by the Americans on Corregidor."

Eisner 29 – Japan began to imprison civilians in Manila. Freed Japanese POWs.

Scott 62-63 – Civilians rounded up, interned at Santo Tomas University. Plundering of homes, cars, trucks, warehouses, food, luxury department stores. Installed puppet government. Took over newspapers, radio stations.

94 Sloan 97-99 – MacArthur's visit draws mixed reviews, Homma drops leaflets. Text. Used leaflets as toilet paper.

Perry 122 – The day of his visit, Jan. 10, Japan began the battle for Bataan.

Eisner 20-21 – MacArthur's reaction to Homma's message. Reaction of soldiers.

Montgomery 55 – Soldiers used Japanese leaflets as toilet paper.

95 Roosevelt, Franklin D. "Message of President Roosevelt to the People of the Philippines"

December 29, 1941. Presidential Museum and Library, Philippines Gazette.

https://www.officialgazette.gov.ph/1941/12/29/message-of-president-roosevelt-to-the-people-of-the-philippines/

Hersey 255-257 – Text of speech, reaction, help was coming.

Manchester 239-240 – Interpretation of Roosevelt speech. MacArthur message Jan 15, 1942, promising help is on the way, timing uncertain.

Morton 242 – Just as well that American and Filipino soldiers on Bataan didn't know how bleak the outlook was for help. "At least they could hope that help was on the way. Only General MacArthur and his immediate staff knew the worst."

96 Goralski 201 – Hitler speech, U.S. attack on Gilbert and Marshall Islands.

205-207 – Java Sea battle, second Pearl Harbor attack, shelling of Santa Barbara, California.

97 Morton 257-258 – Hunt for food more important than war for some, everyone hunted at all hours.

98 Morton 387 – MacArthur had promised his troops on January 15, 1942, that help would soon be on the way. Roosevelt's speech , delivered on Washington's birthday at home (Feb. 22) was "a rude blow" to the hopeful soldiers.

Roosevelt "On Progress of the War" Fireside Chat, February 23, 1942.

Franklin D. Roosevelt Presidential Library and Museum http://docs.fdrlibrary.marist.edu/022342.html

Philippines surrounded. Attack on Pearl Harbor wasn't as bad as most said, just too far away to help the Philippines.

99 Morton 76-77 – The fall of Wake and Guam cut the line of communications between Hawaii and the Philippines and left the United States with no Central Pacific base west of Midway, 4,500 miles from Manila. But even before this, on the first day of war, the Japanese attack on Pearl Harbor had destroyed the Battle Force of the Pacific Fleet and nullified all plans to come to the aid of the Philippines.

387 – Roosevelt's Feb. 23 speech making it clear no help was coming. The troops' "hopes received a rude blow."

100 Goralski 205 – Japanese submarine shells California oil refinery.

101 Goralski 205 – Feb. 22, 1942, Roosevelt orders Mac-Arthur to leave the Philippines. Japanese submarine attacks oil refinery near Santa Barbara, California.

Sloan 139-141 – MacArthur's reaction to the order to leave. Hands command to Wainwright.

102 Hersey 5-6 – MacArthur as a symbol.

Underwood, Charles Jr., *Deadline: Captain Charlie's Bataan Diary*, Piscataqua Press, (2013).

87 – like pulling star quarterback.

103 Daws 66 – Press releases. Soldiers choked at the sound of MacArthur's name.

Manchester 230 – Of 142 communiques sent out by Mac-Arthur in the in the first three months of the war, 106 mentioned only one soldier – MacArthur.

232 – "The messages were self-advertising, and hard-sell advertising at that. Americans at home, hungry for heroes, accepted them at their face value, but in retrospect they sound stilted and turgid. They weren't even accurate." They reported the sinking of ships, and that his army was greatly outnumbered. And that Homma had killed himself after failing to succeed by the Tokyo gave him.

Scott 130 – MacArthur spent hours each day crafting statements. But he remained popular at home.

104 Considine, Bob, *MacArthur the Magnificent*, serialized in numerous newspapers in March 1942. Considine, a journalist, is best known for his books *30 Seconds Over Tokyo*, and *The Babe Ruth Story*. He was also the editor for *General Wainwright's Story*.

Pittsburgh Sun Telegram, March 1, 1942

Gaining territory on Bataan

St. Louis Star-Times, March 17, 1942

"MacArthur's Bataan Stand Ranks with Alamo, Thermopylae"

Richmond Times-Dispatch, March 13, 1942 (and other newspapers) Commentary by Hugh S. Johnson impregnable defenses on Bataan "Youth Not Assurance of Great General"

Fortification of Bataan and impregnable mountain defenses.

105 Groom 174-175 – Japan temporarily halted its offense, followed by an ominous calm.

Toland 286, 303, 366 – Difficult landing at Lingayen Gulf. Homma got a telegram from Tojo in late January criticizing him for not having won the battle. Brought home, retired in disgrace. Homma had spent eight years with British.

325 – Mid-February. Bataan quiet. Men on one-third rations, killed all the horses and mules, no fodder for them. Bataan one of the most-malaria-infested places in the world, and the supply of quinine almost gone. Despite rumors of reinforcements, hope was gone.

Morton 347 – By not conquering the Philippines by the assigned deadline, Homma failed miserably.

Perry 148 – General sent to inspect found Homma's army in chaos. Asked for and got reinforcements.

176 – Homma returned to Japan.

Sides, Hampton. *Ghost Soldiers: The Epic Account of World War II's Greatest Rescue Mission*. Anchor Books. 2002.

58-60 – Homma, tall, Western-educated.

Manchester 201 – Homma had been ordered to conquer the Philippines in 60 days.

106 Morton 298-299 – Battle of the Points, Jan. 23 to Feb. 11, 1942, the defense of the west coast of Bataan. Some members of the provisional infantry took part in the battle.

324 – U.S. and Filipino troops wiped out an entire Japanese battalions. The battle ended with an allied victory by mid-February.

107 Schaefer 32-33 – Message to Wainwright in empty beer cans.

Morton 418 – Beer can message from Homma, failing to surrender would be disastrous. An ominous warning that if Wainwright did not reply by noon March 22, Homma would consider himself "at liberty to take any action whatsoever."

108 Wainwright 52-53 – Bataan a hopeless hell. Malaria hung over them.

109 Hersey 98-99 – characterized Japan's fighting skills after some said Japanese soldiers were inept.

Perry 136-139 – Japan conquered one fourth of the world's surface and was free to roam the Pacific. Skills of Japanese soldiers compared to skills of U.S. soldiers.

Manchester 234 - The Japanese empire stretched 5,000 miles in every direction from Tokyo. Controlled a seventh of the world, that it was mostly water made it easier to defend.

110 Goralski 209 – Homma resumes attack the same week.

Morton 401-402 – Washington learns how many people are on Bataan, not enough rations.

Sloan 50 – Congress and Roosevelt were partly to blame for lack of resources in Philippines. Not enough resources, the Europe first policy.

Chapter Seven

111 Morton 421 – April 3, anniversary of the death of legendary emperor, Jimmu, also Good Friday.

442 – Collapse of army into thin air.

Dooley 77 –Visit to Bataan, defeat written on their faces.

112 Toland 329 – Combat effectiveness.

Morton 456 – "The only alternative remaining to King if he followed Wainwright's orders was to accept the wholesale slaughter of his men without achieving any military advantage. Under the circumstances, it was almost inevitable that he would disobey his orders." Roosevelt had ordered King not to surrender.

Dooley 80 – MacArthur ordered King not to surrender.

113 Wainwright 67-69 – 3 a.m. call, King did not mention that surrender was underway. Wainwright later wrote about April 9 that a silent battlefield was worse than one that was shaking.

Sloan 167-168 King expected to be court martialed.

Toland 332-333 – King removed his staff from decision making, said he expected to be court martialed if he survived. Earthquake.

Archive Record, Major General Edward P. King Jr., American Defenders of Bataan and Corregidor Memorial Society.

King spent 3½ years as a POW and retired in 1945. He was not court martialed. He died in 1958. https://philippinedefenders.pastperfectonline.com/archive/86CD8FC4-6D77-45D2-A008-167063116554

King, Edward P., MG, Together We Served https://army.togetherweserved.com/army/servlet/tws.webapp.WebApp?cmd=ShadowBoxProfile&type=Assignment-Ext&ID=404590

Dooley 83 – Knows the real person King is, with a minimum of aid Bataan could have been held.

114 Hardee/Blazich, preface xvii - "Outmanned, outgunned, and unsupplied, the Americans and Filipinos were overwhelmed by the Japanese and forced to surrender on April 9, 1942. It remains the largest surrender in American history."

David Hardee spent 34 months in captivity. What followed "death and misery unprecedented in the American military experience." 78,100 surrendered, including 11,796 Americans on Bataan, 25,580 captured in the Philippines mid-1942. Of those 10,650 would die in captivity.

Tenney, Lester I., *My Hitch in Hell: The Bataan Death March.* Potomac Books, 2010

197 – 12,000 of those surrendered were Americans.

Sloan 177-178 – Japanese accept surrender. Claimed they were not barbarians, but a cold-blooded orgy followed.

Toland 334 – Roosevelt withdrew order to attack.

115 Manchester 201 – Homma had been ordered to conquer the Philippines in 60 days.

235 – On Feb. 8, Homma's deadline, he asked for more troops.

Daws, Gavin. *Prisoners of the Japanese: POWs of World War II in the Pacific.* William Morrow and Company Inc. 1994

72 – description of ammunition being blown up.

116 Schaller 72-73 – MacArthur's press releases.

Scott 130 – Press releases, many erroneous, many mentioning only him.

203 – press releases about victory issued before the battle of Manila had begun.

117 Miller 197 – Food seized and destroyed.

118 Allen, Oliver "Red," as told to Mildred Allen. *Abandoned on Bataan: One Man's story of Survival.* Crimson Horse Entertainment and Publishing. 2002

64-65 – Turned in weapons, other objects, but turning in food was the worst.

119 Knox 114-115 – fear as Japanese soldiers shouted orders they could not understood.

120 Sides 46 – King had trucks and gasoline is reserve.

57 – Reasonable march, 10 miles per day with food and water.

Sloan 298, 316-319 – Not until January 1944 did Americans learn of the march, through Dyess' story.

121 Sides 90 – Estimates of deaths on the march vary widely, especially for Filipino soldiers, but 750 U.S. and 5,000 Filipinos are the best estimates.

Toland 343 – No one knows, but of 7,000 to 10,000 deaths on the march 2,300 were Americas.

Hardee/Blazich xviii – Death march killed between 600 and 650 Americans, 10,000 Filipinos. Another 4,100 Americans died in POW camps by the end of 1942. Thousands more died on hellships.

122 Hardee 55 – A group of Japanese soldiers took his canteen and between 30 and 50 from other soldiers. He bought one from someone who had two for 3 pesos. Guards also took helmets.

123 Dyess 18 – The wanton murder of a captain.

55 – He wanted to kill Japanese soldiers who looted, beat, and killed POWs caught with Japanese-made items or money.

19 – Japanese soldiers rolled an unconscious body under a truck.

124 Coleman 72 – Thought he had slipped in the mud, it was human flesh, pulverized.

Knox 121– Quoting a witness who saw tanks run over a live soldier – uniform embedded in the cobblestones.

Underwood 102 - Day of the march, near Orani, a tank came out on road and crushed four prisoners.

125 Sloan 201 – 2,000 soldiers escaped from Bataan to Corregidor, Japan continued to fire at the island.

Morton 471 – 2,000 escaped to Corregidor, most in small boats and barges.

Wainwright 79-80 – Japan occupied high spots on Bataan, so they could aim at artillery on Corregidor.

Norman, E. 85 – 3,000 soldiers evacuates.

4-5 – Number of nurses evacuated.

126 Groom 177 – Assumptions versus reality of the march. Too many witness statements to doubt the cruelty.

127 Tenney 50 – guards were getting revenge for their comrades.

Michno, Gregory F., *Death on the Hellships: Prisoners at Sea in the Pacific War*, Naval Institute Press, 2001.

132-133 – Under Japan's bushido code, death in war was noble, suicide wasn't. Victory would be sought by any means.

Daws 18 – Old bushido code didn't carry over to guards treatment of POWs: "In the eyes of the Japanese, white men who allowed themselves to be captured war were despicable. They deserved to die."

Norman and Norman 81-82 – The bushido code was derived from a myth.

Dallek 193 – 1938 Japan wanted a new world order.

Toland 507 – Greater Asia Co-Prosperity Sphere.

128 Sides 57 – Homma's plan was for soldiers who were able to march 10 miles per day with food and water. Vehicles would transport those unable to march. Hospitals would be set up along the route. Based on wildly inaccurate intelligence reports, Japan expected 25,000 POWs. Homma asked for better data, and the estimate was 40,000, still way off.

Perry 357 – Homma tried and executed in 1946.

Toland 343 – Japanese propaganda that the march was humane, blamed the U.S. for soldiers' poor conditions.

129 Toland 335 – Expected to capture 25,000 healthy soldiers with their own food supplies.

Sloan 133 – Japan acquired more supplies during the lull. Initially, Japanese soldiers also had food shortages.

Morton 412 - Even by Japanese standards the lot of the soldier on Bataan was not an enviable one. Certainly he was not well fed. Rice supply had run low by February.

Underwood 84 – Captured Japanese soldiers said they were short of food.

Dyess 53 – U.S. soldiers with Japanese goods or money beaten or killed.

Knox 118 - In addition to underestimating the number POWs, Japan was behind in planning how to transport and provide for them. It also assumed the Americans and Filipinos would be healthy.

130 Tenney 55 – Filipinos throwing food at marchers, then shot.

Dyess 70-71 – Filipinos throwing food and being punished.

Hardee 58 – Allowed to buy food, a guard gave him some.

131 Sloan 184-185 – Brutality escalated into a frenzy, descriptions of what men saw.

132 Daws 77 – At stops, made POWs stay in sun, even if there was a shady area nearby with water. Known as "the sun treatment." As men began to faint, they herded them back onto the road and made them march double time.

133 Knox 138 – Others reported a sense of confusion.

Coleman 206 – Many on the march reported they felt "foggy."

Dyess 66 – Men without food and water: "They were like Zombies – the walking dead of the Caribbean."

Underwood 102 – Those who failed to get up "used like pincushions for Japanese bayonets." He saw one Filipino POW beheaded when he didn't get up quickly enough.

104 – By the fourth day, many had what he called "a walking coma."

134 Groom 180 – Dead bodies along the road.

Hardee 49, 53, 60 – Dead bodies and the stench.

135 Dyess 74 – POWs endured a three-hour ride in over-crowded boxcars.

Wainwright 80 – The box cars went 20 miles.

136 Hardee 60 – More bodies along the route.

137 Ty did not name Betis as the place where he was living, but researcher Guill Ramos has interviewed members of the

Jose family, Ty's main benefactors, who said that is most likely where the American escapees were living. Family members often went back and forth between the farm and Guagua, and Ty never mentioned that it was a long journey.

Chapter Eight

138 Martin and Stephenson 144 – "After the Japanese conquest of the Philippines, Filipinos often hid Americans and looked after them." Some jointly carried out guerilla operations.

139 Dooley 87 – Japan unleased a terrific bombardment on Hirohito's birthday, a few officers and nurses evacuated from Corregidor.

88 – Nurses, others evacuated May 3.

Norman, E. 29 – The remaining 77 nurses were taken to Santo Tomas University in Bataan, where they were imprisoned with civilians.

Jacoby, Annalee. "Bataan Nurses: Nurses Under Fire in the Philippines: April 1942." As told to Annalee Jacoby by Willa L. Hook and Juanita Redmond.

Reporting World War II: American Journalism 1838-1946. The Library of America. 1995. Advisory board: Samuel Hynes, Anne Matthews, Nancy Caldwell Sorel, Roger J. Spiller.

132 – Nurses called on short notice then flown out of Corregidor.

Goralski 217 – U.S. and Filipino soldiers told to give up arms by May 10, but formal resistance lasted another month.

215 – Corregidor surrendered. Wainwright: "We are beaten..." The Battle of the Coral Sea.

Philippine Trade Act of 1945. Hearings Before the Committee on Ways and Means. October and November 1945 and February and March 1946.

296 - Wainwright quote, "We are beaten ..." https://tinyurl.com/y44k7mkr

Wainwright 88-89 – During April evacuated military personnel and a few reporters.

140 "Keep 'Em Flying" Vintage WWII U.S. Army Air Corps Recruitment Poster, 1942

Army Air Corps recruiting poster https://www.great-republic.com/products/keep-em-flying-vintage-wwii-u-s-army-air-corps-recruitment-poster-1942

141 Wainwright 211 – Wainwright kept track of his solitaire, games, percent won.

142 Koyfman, Steph, "What Language Is Spoken in the Philippines?" Babbel, July 31, 2019

https://www.babbel.com/en/magazine/what-language-is-spoken-in-the-philippines

183 languages. Filipino and English are official. Spanish also widely spoken. Kapampangan another major dialect. (Ramos, Guill – Kapampangan is widely spoken in Pampanga.)

Philippines, One Page Summary, CIA World Fact Book

Eight major languages. https://www.cia.gov/the-world-factbook/static/48cea11edf6a49859800dd0066ca9e2f/RP-summary.pdf

Ramos, Gull. Researcher based in Manila.

Ty wrote that *busig-nako* means "I am satisfied." He was close: "busog na ako" is a Tagalog phrase that means "I'm already full," usually said after a meal. In Kapampangan the same phrase is: "Mabsi naku."

143 Hardee 135-136 – Hunger, food, recipes.

144 "1940s: NADA and WWII," National Association of Car Dealers." https://www.nada.org/1940s/

"Automobiles in the Postwar Economy," CQ Researcher. https://library.cqpress.com/cqresearcher/document.php?id=cqresrre1945082100

145 Knox 313-333 – Huks and other guerilla groups. An oral history with some American guerillas, takes up a chapter in his book.

Sloan 286-292 – Huks and other guerilla groups, including Thorp and Col. Edwin Ramsey.

146 Ty's comments about politicians and taxes appear only in the transcript at the National Archives. They do not appear in the transcript his parents had made. It's possible that by the time Hans and Shy got the diary, those pages were unreadable because of mold. It is also possible they didn't want his angry writing in copies they intended to share with others after the war was over.

Jones 55 – Arden R. Boellner expressed similar sentiments, writing that it should be a duty of politicians to go into the field with the army, somewhere "where nice talk doesn't help." He and Ty arrived in Manila on the same ship, but there is no evidence they met. Boellner spent most of his time on Cebu and Mindanao. He was killed the when the Oryoku Maru was bombed in Manila Harbor.

147 Daws 100-101 – "In the Philippines, some helpful Filipinos might have been ready to harbor Americans." It was less likely elsewhere in the Pacific. The number of escapes overall was minute. White POWs imprisoned in their skins, almost all escapees died, as did many in camps.

Knox 271-272 – Others who contemplated escape said they did not know the area or language, didn't want to be a burden on local residents and decided escape too risky. Lear and Tussing.

Heisinger, Duane. *Father Found: Life and Death as a Prisoner of the Japanese in World War II* (2003) Xulon Press

191-192 – Escape practical for only a few. Poor health, security at camps. "Men did slip away in the hours around the time of the surrender, amazingly even a few during the March itself." Some eventually reached safety, others were intercepted. A few reached Australia, the target for most. Some harbored by Filipino families. Some joined guerilla groups.

Lawton, Manny. *Some Survived: From a Soldier Who Was There ... The Shocking True Story of The Bataan Death March.* Warner Books. 1984.

9 – He thought about escape: "For a blonde Caucasian to melt into the Oriental populace and be unnoticed, I reasoned, would be impossible." Hiding in the jungle, maybe. But he would need quinine and none was available. He decided to stay put.

148 Goralski 212 – Doolittle bombing raid over Japan.

Toland 348-354 – Doolittle raid.

149 Goralski 217 – Burman fell to Japan, May 15, 1941.

218-219 – U.S. won the decisive Battle of Midway, June 4-6. U.S. intelligence meant the fleet was not diverted from Midway when Japan attacked two Aleutian Islands in Alaska territory. Japan was no longer infected with "victory disease," as it was in the first six months of the war.

Devastating RAF attack on Cologne, Germany: 12,000 fires, 486 killed, 5,027 injured, 59,100 left homeless, 18,432 buildings destroyed, 9,516 building heavily damaged. Enough factories damaged that production was halted for three months.

222 – Japan shelled Fort Stevens, Oregon, June 22, 19424. No casualties or damage.

Groom 216 – U.S. decoded Japan's plan for Alaska diversion. Adm. Chester Nimitz knew Japan had assembled a fleet of about 200 ships. Biggest so far.

218 – U.S. figured out attack at Aleutians would be June 2 and attack on Midway June 3. Japan's fleet was much bigger than the U.S. fleet.

220 – Japan assumed attack on Aleutians would draw some if not all of the U.S. fleet to Alaska. Also that U.S. would respond to the Japanese attack on Midway, not anticipate it.

225-231 – Midway Battle. "For the Japanese, 'Victory Disease' had set in to the point that most believed themselves invincible."

Toland 376-292 – Battle of Midway. Japan concealed information about its Midway defeat from the public at home.

591 – Even after the Midway defeat, Prime Minister Hideki Tojo fails to realize or acknowledge the growing U.S. power in the Pacific and Japan's losses.

Chapter Nine

150 Malaria: Frequently Asked Questions (FAQs), The Centers for Disease Control and Prevention, January 26, 2021. https://www.cdc.gov/malaria/about/faqs.html

Paltzer, Seth, "The Other Foe: The U.S. Army's Fight against Malaria in the Pacific Theater, 1942-45," The National Museum of the United States Army.

https://armyhistory.org/the-other-foe-the-u-s-armys-fight-against-malaria-in-the-pacific-theater-1942-45/

The National World War II Museum, "WWII Post Traumatic Stress," June 27, 2020. https://www.nationalww2museum.org/war/articles/wwii-post-traumatic-stress

151 Goralski 227 – U.S. began the invasion of Guadalcanal and the Solomons Islands, Aug. 7, 1942.

U.S. loss at Savo Island, Aug. 9, 1942.

229 – U.S planes bombed Formosa, Aug. 13, 1942.

237 – Roosevelt called for drafting men ages 18 and 19, Oct. 12, 1942.

Roosevelt signed the biggest tax bill in history, Oct. 21, 1942.

U.S. loss at Savo Island, Aug. 9, 1942.

240 – The U.S. Army announced 800,000 Americans were serving overseas.

257 – All organized resistance ended on Guadalcanal, Feb. 9, 1943.

310 – Last major battle in the Solomons, March 24, 1944.

152 Dyess 97 – He (and Ty)feared he would forget all the table manners he had learned.

153 Schaefer 30 – Major Frank Loyd became a guerilla, hid his diary, and returned to the Philippines after the war to repay his debts.

334-339 – Loyd was taken to Guagua in February 1945 after fighting ceased nearby, to begin his journey home.

154 Field, Eugene, "Little Boy Blue," 1911. https://allpo-etry.com/Little-Boy-Blue

Chapter Ten

155 Jack Dinsdale visit: Jack's younger brother, Roy, and my father met in college, and they all remained friends for the rest of their lives. Our family spent occasional, wonderful weekends at the Dinsdale farm and feedlot. The two brothers' houses were separated by a swimming pool and backed up to a cornfield.

Ganzel, Bill, "The Rise & Fall of the Omaha Stockyards," Ganzel Group, 2007.

https://livinghistoryfarm.org/farminginthe50s/money_14.html

Carter, John E., "The Birth of the South Omaha Stockyards," History Nebraska, 2013

https://history.nebraska.gov/blog/rare-pictures-omaha-stockyards

156 Martin and Stephenson 87 –The original plan for the 27th Bomb Group was to be ready to move to Mindanao on short notice. When only one ship was available on Dec. 29, the smaller 19th Bombardment Group was sent instead.

Morton 156 - The 650 men of the 19th Bombardment Group left Luzon before the end of December to join their planes in Australia. Their comrades in the 24th and 27th Groups were not as fortunate.

Prime, John Andrew, "Our History: Doomed bomb group valiant," Shreveport Times, Feb. 15, 2015.

Tells the story of the few 27th Bomb group members who made it to Australia, and about those who did not. https://www.shreveporttimes.com/story/news/local/ 2015/02/15/history-doomed-bomb-group-valiant/23458429/

157 Minor, Colin. "Filipino Guerilla Resistance to Japanese Invasion in World War II," Legacy, Volume 15, Issue 1, 2015, P 35. Southern Illinois University, Carbondale.

A majority of Filipinos are Catholic. The Moros are Muslims. They live on several southern Philippines islands, including Mindanao. They opposed American control of the Philippines, but assisted U.S. troops during the war because they were more opposed to Japanese control. Moros opposed the U.S. takeover in the early 20th century, but the Muslim group disliked the Japanese more and were successful guerillas in concert with Americans. https://cola.siu.edu/history/undergraduate/legacy/

158 G.I. Bill. History.com Editors, FDR signs G.I. Bill https://www.history.com/this-day-in-history/fdr-signs-g-i-bill

U.S. Department of Defense, 75 Years of the GI Bill: How Transformative It's Been

January 9, 2019 https://www.defense.gov/News/Feature-Stories/story/Article/1727086/75-years-of-the-gi-bill-how-transformative-its-been/

159 Andrews Sisters, "I'll Be With You in Apple Blossom Time," 1942, You Tube. Lyrics and performance. https://www.youtube.com/watch?v=c4gyB-IWU00

160 Rosenfeld, Megan, "Government Girls: World War II's Army of the Potomac," The Washington Post, May 10, 1999.

Her mother was one of 200,000 government girls. The article describes their work and crowded and substandard boarding houses. https://www.washingtonpost.com/wp-srv/local/2000/govgirls0510.htm

161 Goralski 224 – U.S. planes joined RAF planes on July 4, 1942, in first European bombing involving Americans.

162 Einstein, Charles. "When Football Went to War: The seasons of 1942-45 turned the game upside down, creating new juggernauts and decimating some old ones," *Sports*

Illustrated, December 6, 1971. https://vault.si.com/vault/1971/12/06/when-football-went-to-war

Beran, Mike "Red," "Brief History a of University of Nebraska Football," [sic], History of the Nebraska Cornhuskers, Football Letterman's Association. https://huskerfootballletterman.org/history

163 Edmonds 172-173 – Dec. 17, word received that the planes for the 27th were headed for Australia.

241 – In footnote: There were between 6,000 and 7,000 Air Forces members on Bataan. 5,000 were surrendered in April. Some were killed, 650 went to Mindanao, and a few went to Corregidor.

164 "Alfred M. Landon: Politician, oil man. Republican," Kansapedia, Kansas Historical Society. December 2004, updated October 2017

Alf Landon, short biography https://www.kshs.org/kansapedia/alfred-m-landon/12126

"Alfred Landon's Acceptance Speech," Kansapedia, Kansas Historical Society

April 2010, updated December 2017

The text of Landon's speech accepting the Republican nomination to run for president, July 23, 1936. https://www.kshs.org/kansapedia/alfred-landon-s-acceptance-speech/14501

"1936 Electoral College Results." National Archives and Records Administration.

https://www.archives.gov/electoral-college/1936

The American Presidency Project, John Woolley and Gerhard Peters. UC Santa Barbara

https://www.presidency.ucsb.edu/statistics/elections/1936

Even Nebraska, a generally Republican state, voted for Roosevelt in 1932 and 1936.

Dallek 131-132 – Roosevelt carried every state but Maine and Vermont in 1936 election. The largest electoral vote since

1820. No evidence foreign affairs was a big influence. [In 1820, Monroe won re-election without a major opponent.]

Chapter Eleven

165 Goralski 219 – Battle of Midway, June 4-6. The U.S. knew of Japan's plans for a battle at Midway and that the Aleutians invasion was meant to be a distraction. The battle was a decisive American victory and a turning point in the way.

220 – Japan invaded Attu and Kiska, two Aleutian Islands in Alaska territory, June 7, 1942.

277-278 – Japan withdraws its troops from the Aleutians, Aug. 15, 1943.

337 – Japan was defeated on Guam, Aug. 10, 1944.

166 Goralski 203 – On Feb. 12, 1942, Japan occupied Borneo's capital, Banjarmasin, which was the capital of Dutch Borneo and is now part of Indonesia.

414 – Japan's occupation lasted until the end of the war.

167 Nasuti, Guy. "The Forsaken Defenders of Wake Island," Archives Division,

Naval History and Heritage Command, December 2016

Wake surrendered Sept. 4, 1945. https://www.history.navy.mil/browse-by-topic/wars-conflicts-and-operations/world-war-ii/1941/philippines/defenders-of-wake.html

Scott 77-81 – By the end of the war, 3,700 men, women, and children were interred at Santo Tomas University. At Bilibid prison, there were 1,275 military prisoners. There were another 2,600 at other camps, including Cabanatuan.

At Palawan, prisoners were burned alive, shot, or stabbed. Of 150, 11 lived.

168 Goralski 243, 407 – Australia began to regain territory in New Guinea, November 12, 1942. But the battled ended with the end of the war.

242-243 – Allied victory in Casablanca, Nov. 7, 1942.

92 – Spanish dictator Francisco Franco declared neutrality in public, but in private pledged support to the Axis, Sept. 4, 1939. The United States declared neutrality the next day.

135 – Franco and Hitler met in France on Oct. 23, 1940. Hitler refused Franco's demands and Franco began to back away from fighting.

210 - Russia bogged down on the front with Germany.

240-241 – Russia continued to block Germany.

249 – Allied forces reach Buna, New Guinea, but were slowed by heat and malaria, Dec. 5, 1942.

The National World War II Museum: "The Eastern Front." https://www.nationalww2museum.org/war/articles/eastern-front

169 "Research Starters: US Military by the Numbers," The National World War II Museum.

In 1943, more than 9 million Americans were serving in the military, up from 3.9 million in 1942. https://www.nationalww2museum.org/students-teachers/student-resources/research-starters/research-starters-us-military-numbers

The UVA Miller Center. Presidential Speeches

Franklin Roosevelt, September 7, 1942: Fireside Chat 22: On Inflation and Food Prices

https://millercenter.org/the-presidency/presidential-speeches/september-7-1942-fireside-chat-22-inflation-and-food-prices

Resources devoted to the war, including three times as many soldiers sent overseas in World War I.

Columbus Day speech

The UVA Miller Center. Presidential Speeches

Franklin Roosevelt, October 12, 1942: Fireside Chat 23: On the Home Front

"I can say to you that we are getting ahead of our enemies in the battle of production."

"I believe that it will be necessary to lower the present minimum age limit for Selective Service from twenty years down to eighteen." https://millercenter.org/the-presidency/presidential-speeches/october-12-1942-fireside-chat-23-home-front

Manchester 146 –Before entering the war, the U.S. had 16th largest army in the world, 132,069 soldiers. It was smaller than the armies of Portugal and Greece.

170 Morton 295-296 – "Since the Philippine Constabulary had been demobilized with the invasion, the mass of people in the outlying areas was left without adequate police protection. Such a situation soon encouraged the rise of numerous marauding parties which roamed the countryside in search of easy loot and tribute to be taken from the defenseless farmers."

171 Schaefer 206-209 – Thorp and other guerillas executed after sham trials.

172 Goralski 235 – First U.S. jet plane flies in Mojave Desert, Oct. 1, 1942.

Burrows, William E. "A Bell That Didn't Ring: Turns out that jets are like waffles: The U.S. Army Air Forces was tempted to throw its first one away," Air & Space Magazine
September 2005 https://www.airspacemag.com/military-aviation/a-bell-that-didnt-ring-7948421/

Goralski 246 – First nuclear chain reaction, University of Chicago, Dec. 2, 1942.

248-249 – Anthony Eden tells Parliament about German genocide, the United Nations announced crimes against Jews would be prosecuted, Dec. 17, 1942. The United Nations said those who committed crimes against Jews would be prosecuted.

United Nations, Model United Nations, history

On Jan. 1 and 2, 1942, 26 countries fighting Axis powers, signed a document called the Declaration of the United Nations. https://www.un.org/fr/node/44721

Huxen, Keith, "Operation Husky: The Allied Invasion of Sicily," July 12, 2017. The National World War II Museum. https://www.nationalww2museum.org/war/articles/operation-husky-allied-invasion-sicily

173 Bjoring 92-95 – Account of his journey on a hellship to Japan.

Michno 86 – Only two hellships were torpedoed in 1942 of 54, with a death rate 4.5 percent. "It was about to get worse."

International Committee of the Red Cross, "Convention relative to the Treatment of Prisoners of War. Geneva, 27 July 1929." https://ihl-databases.icrc.org/ihl/INTRO/305

Work by POWs was legal if they were paid, worked in decent conditions, and if they were not sick. Work that contributed to the war effort was illegal.

DVA (Department of Veterans' Affairs) (2020), "Burma-Thailand Railway and Hellfire Pass 1942-1943." https://anzac-portal.dva.gov.au/wars-and-missions/burma-thailand-railway-and-hellfire-pass-1942-1943

Eisner 145 – Numbers of POWs in the Pacific and Europe, and death rates.

Goralski 253 – Casa Blanca Conference. Discussing the timing of cross-channel invasion. Roosevelt calls for unconditional surrender, likely prolonging the war. Jan. 14-23, 1943.

Chapter Twelve

174 Underwood 114 – O'Donnell known as "Camp O'Death."

Toland 343 – The number of deaths on the march and who reached O'Donnell are all estimates. Japanese newspaper report. "Brutality to the Japanese soldier was a way of life." No such thing as surrender; death was preferable. Easier to follow orders than take initiative. Some commanders wanted to kill all POWs, an influence for brutality.

Lawton 6 – March hell; O'Donnell a new kind of torment.

Sloan 231 – lack of water, slow spigots. Men fought over water.

175 Daws 159 – Nothing but O'Donnell continued, death rate

Groom – 356 – Camp description, diseases. POWs figured they would all be dead within a year at rate they were dying. Up to 120 men in barracks designed for 40. Still a slight improvement.

Fewer than 9,000 of the original 12,000 American captured got to Cabanatuan. The rest died or were murdered. Those who did arrive were in terrible shape. More water than O'Donnell but polluted. Squads of 10 to prevent escape. Cabanatuan means place of rocks. It was a half-finished training camp.

Reynolds 97 – Crowded barracks.

Daws – 124 – Lack of soap and toothpaste.

Sloan 253 – Jan. 18, 1943, first 24-hour period with no deaths. First Red Cross boxes at Christmas helped, they had Spam, potted meet, corned beef, chocolate, jam cheese, dried milk, and other food.

Groom 364 – Camp life somewhat better with a sewer system. "It remained a hell on earth." Gardens supposedly to feed the prisoners, but Japanese took most of it. When fillings fell out of rotting teeth, dentists melted down coins and hammered them into teeth. Fly swatters made of palm leaves. Prizes for killing the most flies.

Heisinger 244-245 – Latrines, septic system. Lime reduced number of flies and dysentery rate. Set up rain barrels.

Coleman 102 – Stood under eaves during rain for showers.

Daws 118 – Groomed each other like monkeys.

176 Michno 310, 317 – Lengths, dates of journeys, alternate ship names.

Tenney 101 – He worked on a detail to take American tanks and trucks apart for scrap metal.

Miller 242 – Some POWs were cooks.

Lawton 46-47 – Some gathered wood, built roads and airfields construction stevedores in Manila.

Jones 83-84 – Her father acquired some of the parts for a secret radio.

Wainwright 157, 164-165 – His camp jobs.

177 Jones 79 – POWs got their first cigarettes in December 1942, then in January 1943, each POW got two 11-pound Red Cross boxes. Then they were allowed to write postcards home for the first time.

Dyess 128 – He saw men crying when they got their first boxes in 1942.

Reynolds, Robert V., *Of Rice and Men*, The Leicht Press, (1949, 1951).

128 – Red Cross boxes during 1944 at Cabanatuan. "The Jap officials at the camp took what they wanted before turning the goods over to us." Some POWs sold what they didn't want or need to Japanese guards.

Lawton 69 – He got his first Red Cross Box in March 1943. It was like Christmas morning. "But now, as a grown man, hungry, weak, lonely and beginning to feel that my comrades and I had been written off and forgotten, the packages had a profound meaning. ... We knew somebody cared."

Daws 18 – Japanese guard looted Red Cross boxes.

146-147 – In a camp near Shanghai, rare Red Cross boxes. "The Japanese were brazen about looting it; they would take as much as a third, even half." They didn't like the smell of cheese, so didn't take it. Still, the boxes were a help. Spam, which was new to most POWs, was a treat.

178 Daws 128 – If POWs indicated poor health, the card would not go out. "Circling the less sunny options did not improve a man's chance that the Japanese would forward his card." Sometimes cards were sent; sometimes not. Route of incoming mail: Europe, Tehran, Soviet Union, trans-Siberian railroad to Asian coast of Russia; from Vladivostok to Tokyo. From there to the camps, or not. A man with just one letter would read it over and over.

Pearson 124-126 – POWs were careful about choosing which of the four health categories to check.

179 Wainwright 171 – Allowed to write his wife three times, got six of her 300 letters.

180 Michno 133 – There were a few decent commanders, but most were drunks, homicidal, sadists. Survival depended on luck.

Hardee 64 – Standing in the sun, listening to the camp commander's speech.

Reynolds 97-98 – 10-man squads. If one escaped the others would be shot

100-101 – Escapees shot. May or may not have been the one reported by Underwood or Dyess.

Hardee 73 – Ten-man squads were "blood brothers." If one escaped, the other nine would be shot. "Few tested the system."

Underwood 116-117 – Three officers tried to escape. They were tortured. After digging their own graves, two were shot, one beheaded.

Dyess 105-107 – a more detailed account of the same incident.

181 Hardee 73- Protests of violations of the 1929 Geneva Convention were useless. Japan would observe them "*mutatis mutandis,|*" or with necessary modifications. Tojo, prime minister and minister of war, ordered POWs had to work to eat.

Daws 162-163, 165 – Of the vegetables they raised, they got little.

Miller 238 – Once in a while, they got a bite of meat or egg.

Coleman 102 – He counted 82 insects in his rice one day—he ate them for the protein.

Daws 113-114 – Men hated rice, craved, sugar, meat and cigarettes. Dutch POWs were used to rice, but not Americans.

Underwood 114 – A nonsmoker, Underwood traded cigarettes for fruit or medicine.

Daws 146-147 – Guards were "brazen" in looting Red Cross boxes. Men would gamble with items from their Red Cross boxes. Cigarettes were the most desired. They played for food.

311 – Some traded their food for cigarettes, leaving them with one meal a day, "meaning they were going to die." POWs set up a system of IOUs so that didn't happen.

Toland 677 – Before arriving at Cabanatuan, men had already lost 55 pounds on average.

Groom 359 – Everyone knew if things went on as they had, they would all be dead in a year. The sickest from the march and O'Donnell were dead. The death rate leveled off in fall of 1942. some supplies arrived through smuggling from outside.

Hardee/Blazich 238, footnote 12 – June-July, 1942, 1,287 POWs died at Cabanatuan, rising to 2,536 by the end of the year. In 1943, 102 men died. In 1944, two died.

Tenney 110 – Never given shoes or clothes. Men without suffered, so they removed them from dead men.

Daws 261-262 – Some POWs turned into thieves. They would steal unattended items. If a man was about to die, they took his clothes. And if he wasn't dead the next morning, they took his food.

182 Daws 273-274 – Camp inspections were a sham. Camps were dressed up for photographers.

Toland 676 - Japan claimed it was treating POWs well while allies were not. "Americans were depicted as 'enjoying life at the various prisoner camps.'"

183 Scott 77 – Number of POWs at Bilibid.

177-179 – Bilibid POWs liberated in February 1945.

184 Sloan 268-269 – U.S. regaining territory, and naval superiority.

Groom 339 – Guadalcanal victory can't be overstated. Convinced Americans Japan could be defeated.

185 Toland 505-506 – Japan's Army and navy fought over resources. Commanders in conquered territories weren't developing resources fast enough. Manufacturing and GDP was rising much more in the U.S. than in Japan. The emperor asked Tojo if Japan could win a battle against the enemy somewhere.

Michno 84-85 – Japanese system of supplying troops abroad began to falter, and with no sweeping territorial gains, it because apparent, Japan could be beaten.

186 Goralski 352 – MacArthur wades ashore after two and a half years, Oct. 20, 1944. (He did not repeat this for the cameras, despite rumors that he did, but he later waded ashore at three other beaches.)

353 – Leyte, "greatest battle in the history of naval warfare." Listing of lost ships. Deep decline in Japan's navy as an effective battle force.

Perry 284-294 – Leyte Gulf battle. Why MacArthur ended up wadding ashore. There were too many landing craft and other boats on the beach, and the harbor master said no more could fit.

Toland 642-643 – Leyte Gulf, the greatest array of guns afloat. Beginning of kamikaze attacks.

648 – Japanese losses at Leyte meant its navy would play no more thana minor role for the rest of the war.

Norwich University Online, Military History, "The Largest Naval Sea Battles in Military History." October 20, 2020 https://online.norwich.edu/academic-programs/resources/largest-naval-sea-battles-military-history

Stewart, Sidney, *Give Us This Day: The Powerful True Narrative of the Bataan Death March*, Popular Library, 1956,1958.

105 – Could see bombings at Cabanatuan in September. Taken to Bilibid in October, where they saw more U.S. planes.

187 Scott 72 – U.S. began bombing Manila, Sept. 21, 1944.

422 – The brutal battle caused the deaths of 100,000 civilians, 16,665 Japanese soldiers, and 1,010 American soldiers.

Michno 243 – The bombings brough hope to POWs.

Hardee 182 – Bombings were the sweetest music. They had a secret radio and heard news.

Daws 288 – Sept. 21, 1944, POWs saw American planes. "Suddenly the whole world looked different."

Lawton 108 – POWs still at Cabanatuan were overjoyed when they saw American planes headed to Manila. It was proof the allies were winning.

149-150 – A few weeks later, at Bilibid, they realized they were being taken to Japan; morale sank to lowest point.

188 Tenney 114 – He realized he was being sent to Japan to work, against the Geneva Convention.

126 – He found out when he got there that he would work in a coal mine.

Sloan 284 – The exact number of deaths will never be known.

Michno 36-38 – Tells of the ship's voyage, repairs. The POWs went to a coal mine.

317 – 126,064 POWs on hellships over the course of the war, 21,039 deaths. 156 voyages by 134 ships – example of double counting is that the *Brazil Maru* and the *Enoura Maru* both carried POWs from the *Oryoku Maru*.

150 – "Nineteen forty-four would be Hell Year."

310 – Toko Maru details.

Daws 286 – "Nineteen forty-three was worse than 1942, and 1944 worse again. The worst single time for a POW to be on the water was September 1944." Lists ships several attacks and deaths v. survivors.

189 Michno 258 – In the rush to evacuate POWs, 1,619 men were put on *Oryoku Maru*, Dec.13, 1944.

Jones 88 – POWs were transferred from Davao to Bilibid before boarding a hellship.

90 – Journey from Davao to Bilibid, then to Cabanatuan.

Hardee 154-156 – In May 1944, he was preparing to leave Davao (POW camp also called Dapecol, for Davao Penal Colony) to Manila in terrible conditions.

Kokjer, Emerson. Letter to Sigma Chi, June 15, 1945, returning a fraternity gold star representing Ty's death and sent to his parents. The letter says they had word that Ty was alive in December 1944 and put on a ship. The fraternity apologized and asked that the gold star be returned.

190 Sides 18 – Jan. 9, 1945, U.S. troops land at Lingayen Gulf.

20 – 500 sickest POWs remained at Cabanatuan, down from 8,000.

21 – There were fears Japanese soldiers would kill the remaining POWs.

270-282 – Description of the raid.

The entire book reconstructs the planning and execution of the successful raid and rescue. These pages describe things from the time Americans blasted the camp gate open.

323-326 – The sickest POWs were sent to San Francisco by plane; the rest went by ship, leaving the Philippines in mid-February. After a month on the ship, most had gained considerable weight.

329 – The ship arrived in San Francisco March 8, 1945.

191 Erickson, James W., *Oryoku Maru* Roster

https://www.west-point.org/family/japanese-pow/Erickson-CSV.htm

Zobel, James W., MacArthur Memorial. Email July 8, 2019, confirms Ty was on the *Oryoku Maru*.

Scott 174 – 810 U.S. soldiers too sick to go to Japan remained at Bilibid.

Lawrence, Kerri, "Archives Recalls Fire That Claimed Millions of Military Personnel Files," National Archives News, July 23, 2018. https://www.archives.gov/news/articles/archives-recalls-fire

A fire in July 1973 destroyed between 16 million and 18 million military records in a six-story warehouse in St. Louis. The fire, which burned for 22 hours, destroyed 80 percent of records of those discharged from the military, including by death, from November 1912 to January 1, 1960. After the fire, the National Archives collected records from other sources that can be used to reconstruct some basic service information. Because of that, I have copies of Ty's enlistment and discharge papers, including the cause and location of his death, but other records may have been destroyed.

War Crimes indictment information: General Headquarters, Supreme Commander for Allied Powers, Legal Section, File No. 014.13, Public Relations Informational Summary No. 15 Feb 47

Subject: U.S. vs Junsaburo Toshino, Shusuke Wada, Kazutane Ahira, Shin Kajiyama, Suketoshi Tanoue, Jiro Ueda, Hisao Yoshido. The document contains a detailed account of the journey of the Oryoku Maru and the subsequent trucks, trains, and ships that took the POWs to Japan. It includes 26 pages typed single-spaced on legal paper (8.5 inches by 14 inches) listing every prisoner who started on the ship, with asterisks next to the names of the few who survived.

Prisoners list p. 17

192 Clarion-Register, Jackson Mississippi, Feb. 27, 1995

Obituary for Edward Blaine Claypool, of Jackson, Mississippi. He married in 1947, got out of the Marine Corps, had a job in Washington, and was a postmaster. He was active in the Disabled American Veterans and the Veterans of Foreign Wars. He died in 1995 at age 75, in Clinton, Mississippi.

Heisinger 419 – Some POWs left messages at Bilibid before getting on ships. "For my of these men, their letters, their notes became the final testaments to their families of these difficult times."

193 Michno 306 –Prison camps were bad, hellships were worse.

296-301 – Comparison to slave ships.

194 Norman and Norman – 318-319, 322-323, 325-326 – Long shifts in coal mines, covered with coal dust, which they also inhaled, punishment.

195 Michno 191-193 – All things considered, the *Notu Maru*, was a decent trip.

196 Pearson 136-137 – Preparations for POWs heading to the ship, air raid the morning of Dec. 13, 1944.

Michno 302 – Wada, cruel interpreter, later charged with war crimes.

Pearson 1-3 – Men boarding the ship.

Michno 258 – Men boarding the ship with others, and MacArthur's Packard.

294-295 – Unmarked ships in convoys, many carrying troops and other cargo in addition to POWs, meant targeting ships that likely had POWs aboard because the ships and cargo could do more harm to allies.

Pearson and Michno report slightly different numbers of Japanese civilians, guards, rescued Japanese sailors, and crew member on the ship.

Sloan 274-276 – Loading of ship, description of what happened the first night, 4,000 square feet, 50 deaths. Not all got the rice and seaweed soup lowered into the hold in the evening.

Michno 258-260 - Loading the ship, heat crowding.

Lawton 155-159 – "We felt as if we were trapped in an oven." Loading just the middle hold took 90 minutes. Wada threatened to shoot into the hold if they did not stop shouting for water and air. "Eerie rantings of maniacs could be heard from men crazed by the heat and frustration of it all."

Weller, George. *Cruise of Death*, (1945, Chicago Daily News. Kindle edition, 2015)

57 – "This is the story of the cruise of death. It is the story of 49 days of savagery and tragedy unequaled in the war in the Pacific, of a Japanese-made hell from which only approximately 300 Americans from more than 1,600 emerged alive."

197 Bisno, Adam, Ph.D., *The Japanese "Hell Ships" of World War II*, Naval History and Heritage Command November 2019 https://www.history.navy.mil/browse-by-topic/wars-conflicts-and-operations/world-war-ii/1944/oryoku-maru.html

198 Sloan 278-280 – POWs felt some relief in the water, cleaning filth. But they were being shot at, and some drowned. Returning bombers waved off. POWs were headed to concrete area surrounded by chicken wire.

Michno 258-261 –As breakfast was lowered into the hold, ship was bombed (some say 7 a.m., he says a 8 a.m.). POWs forced to swim ashore, 300 yards, sometimes under gunfire. The next day, returning U.S. planes saw the POWs and flew away. One plane wagged its wings. 1,333 surviving POWs headed to an old tennis court.

Pearson142-143 – "On one hand, the prisoners wanted to cheer their ace pilots on. If the Japanese were all killed, the prisoners could be freed. On the other hand, the firepower necessary to kill the Imperial forces on board would also most likely kill them." Civilians and Japanese soldiers on the upper decks were hit hard.

Leif R. Kloster, obituary. https://www.findagrave.com/memorial/123968009/leif-richard-kloster

http://iagenweb.org/boards/wright/obituaries/index.cgi?read=599271

Heisinger 467-468 – The account of planes wagging their wings and flying away has been widely reported. But Heisinger disputes it, based on a review of after-action reports filed by pilots

199 Michno 292-296 – Most deaths were caused by friendly fire—American planes and submarines attacking Japanese ships.

200 Pearson 153-154, 161-162 – Conditions on tennis court, food, trucks. Estel Myers optimism.

166 – 15 sickest men sent away.

Michno 261-262 – 1,333 POWs remained alive. Tennis court conditions. Trucks Dec. 20, 21 to San Fernando. Fifteen sickest were chosen to be sent to Manila. Instead they were killed.

Heisinger 489 – POWs told to send 15, even though 200 were sick. The rest of the men thought the 15 were lucky, did not discover their fate until after the war.

201 Authors Judith Pearson and Gregory F. Michno report slightly different numbers of POW deaths

Pearson 164-167 – POWs got a few Red Cross boxes, which included drugs. Men were loaded onto box cars, six feet wide, 26 feet long. Injured men were put on the roof to deter U.S. bombers from targeting the train. 150 men per car, just over 1.3 square feet per person, plus 50 on the roof. [these number add up to 1,400 if there were seven rail cars, so either there were fewer o the roof or slightly fewer per car.] The men hoped they were going to Cabanatuan or Bilibid. It got cold at night for those on top, and in the cars, men passed out from lack of air. No one died. After 16 hours, arrived at San Fernando La Union at the northern end of Lingayen Gulf. Ate hibiscus leaves and flowers.

170 – Ocean dip. One canteen of water per 20 men, four tablespoons each, every 90 minutes. Ships arrived, with soldiers, horses, supplies.

173-176 – Approximately 65 prisoners had died by Dec. 26, leaving 1,234. In the night, POWs were ordered to march to the dock. Loaded onto barges with sudden urgency. POWs suspected the guards had learned something. The America fleet had left Leyte and was steaming north.

177 – It was sunrise when the POWs were all aboard. They left even as the last POWs were being forced below.

184 – Dec. 31, ships in Formosa Straits. Dead of winter, very cold. "The men had been without food for so long their hunger pain became just one more ache in an ever-lengthening list if ailments. But the lack of water was the death knell for many." Occasionally got a quarter canteen cup per day. Down to 1,184.

Michno 261-262 – Sickest men murdered. The rest were taken to Lingayen Gulf, 1,070 on *Enoura Maru*, 236 *Brazil Maru*. By New Year's Eve, 16 deaths on Enoura, five on Brazil. Total 1,306.

Scott 76 – MacArthur and a fleet of 818 ships began coming ashore at Lingayen Gulf, January 9, 1945.

Hardee 163 – Two weeks after the ships left the U.S. Sixth Army landed 175,000 soldiers there.

202 Lt. L.W. Johnson. From Oryoku Maru roster: Johnson, Lycurgus W. He was a second lieutenant in the Army Air Forces, 3rd Pursuit Squadron. He was imprisoned in Japan, then Korea. After the war he returned home to Denver, Colorado.

https://www.west-point.org/family/japanese-pow/Erickson-CSV.htm

203 Reports about Alvan Ose's death differ. It is possible his body was on board the *Oryoku Maru* when it sank. A grave marker in Story City, Iowa, and the tablet honoring him at Manila American Cemetery carry different dates of death. The War Crimes document lists his hometown as unknown.

Toland 712-713 – Carnage in the forward hold of the Enoura Maru when it was bombed.

American Battle Monuments Commission, Manila American Cemetery, Walls of the Missing.

Alvan Ose, listed as missing in action, likely date of death, January 9, 1945.

https://www.abmc.gov/decedent-search/ose%3Dalvan

In his Iowa hometown, a plaque in the cemetery and an entry on the website Museums of Story City both list his date of death as December 15, 1944, the day the Oryoku was sunk with bodies on board. http://www.storycityhistory.org/ose-alvan-s.html

https://www.findagrave.com/memorial/98345457/alvan-stanley-ose

Pearson 187-191 – Jan. 8, all men on *Brazil Maru* on deck for rice. Men warned they would be beheaded if they took any of the sugar. All transferred to *Enoura Maru*, 500 into forward hold. 8 a.m. on Jan. 9, bombing, with one into forward hold of the *Enoura*. 40 died in middle hold. Half in forward hold, 250, died and another 200 were wounded.

198-199 – MacArthur and fleet ready to land at Lingayen Gulf, Jan. 8, 1945. All POWs now on the *Brazil Maru*. The ship sailed Jan. 13. [Pearson may be using dates as of U.S., Michno in Japan or P.I.]

Michno 263-264 – Jan. 6, POWs consolidated onto the *Enoura Maru*, cargo loaded on Jan. 8, including sugar, some of which made its way to POWs. Bombed Jan. 9, killing 300 in the forward hold. Others killed in middle hold, when the hatch cover dropped. Bodies removed Jan.12, help provided to wounded men by other POWs. All the bodies were taken ashore and burned in a mass grave.

Toland 712-713 – "The carnage was indescribable. More than half of the five hundred men in the hold were killed outright. For the wounded, shrieking in pain, there was no medicine, no dressings. Nor was there any answer from topside to pleas for help."

Enoura Maru Hellship Memorial Stone Dedication booklet, published by the American Defenders of Bataan & Corregidor Memoria Society, states that the bodies were buried, not burned.

Johnson, who returned home after the war was over, was the subject of a story in the Fort Collins *Coloradoan* newspaper on September 28, 1945.

204 Sloan 282-284 – *Enoura Maru, Brazil Maru* in Lingayen Gulf, with 1,035 POWs still alive. "After the long delay, their captors suddenly were in one hellacious hurry to embark." The ships left even as POWS were still climbing aboard from barges. Jan. 9 at Takao, quotes Lawton about bomb. Now 930 remained alive. In the next month, 652 died. Guards would trade a cup of water for a gold ring. Extremely cold weather, *Brazil Maru* decks coated with ice. Arrived at Moji, Jan. 30, 1945, with 425 alive. Another 161 would be dead within a month, leaving 264 alive from the original 1,619.

Pearson 203-204 – Everyone had dysentery. Buckets used for toilets overflowed. First two days after Jan. 13, no POWs got any water from the ship's full tanks. Hold got cold with hatch open for air. Got salty water to drink on Jan. 15, one canteen cup for eight men, and rice, one canteen cup per four men – each twice a day. Deaths mounted.

212-214 – Guards sold food, water, cigarettes. A few POWs still had wedding or West Point rings. A good solid gold ring could bring two to five canteens of water. Traded shoes for cans of salmon or tomatoes. Caught snow coming through the hatch for water. Deck coated with ice.

217 – Myers description – more like feral beasts than men. Michno 265-266 – number alive when ship reached Moji, only 1,300 POWs still in Philippines left to rescue.

205 Michno 309-317 – complete list of hellship journeys, including the length.

206 Michno 72-73 – Summary of rarely followed order to make sure POWs on ships arrived ready to work.

Pearson 220-221 – *Brazil Maru* arrives in Moji. Japanese military board the ship, stunned at wretched conditions. They had expected 1,000 healthy men ready to work. They sprayed the POWs with disinfectant.

225 – Men got a little clothing and some shoes. But there wasn't enough. Marched through streets, children spat at them. Taken to unheated building, some blankets and overcoats.

Daws 299 – why was the *Oryoku* the worst in terms on death (except for sunk ships). Worse conditions onboard, POWs in worse conditions. Yet no officer was able to take command and stop the killings. Heat, low oxygen levels.

Weller 776 - "You could see that Moji officials were taken aback."

207 Stewart 174-175 – in the morning, separated sick and wounded from the rest to go to a hospital. Old dilapidated wooden building. Put into two rooms. Vacant, little light, damp, cold floors. Counted 49 men, assumed there were 51 in the other room.

Norman and Norman – 318-319, 322-323, 325-326 – Long shifts in coal mines, covered with coal dust, which they also inhaled, punishment.

208 Pearson 226 – Began dividing men into three groups. "The first group divided out was the 'hospital group,' those who were determined to be too infirm to begin work right away, if indeed they ever would at all." The 300 remaining "well" POWs divided into two groups. Ambulances came for hospital group. Myers wondered why no corpsmen were sent with them. "The hospital group of men were in such dire condition, they were never going to recover. The Japanese probably figured there was little point in wasting able-bodied men by sending them out with the doomed. Myers' conjecture proved to be correct." Seventy-six died at the hospital.

Weller 776 - At Moji, local officials asked to see the POWs' senior officer. When they saw him, filthy, hair hanging down, standing between slop buckets and dead bodies. "You could see that Moji officials were taken aback." Between 425 and 435 still alive. Taken to an auditorium, given rice. Volunteers took extremely ill to the hospital. Others taken to work camps.

209 Stewart 174-177 – Being taken to the hospital, food, people dying.

182-185 – Rescued by British POWs turned over to Americans and taken to a POW camp in Manchuria in April 1945, where there was medical care.

210 Michno 306 – Quoting a POW, love didn't keep you alive, hate did. When men gave up, they died.

211 Toll, Ian W. "The Atomic Bombings at 75," The National World War II Museum.

August 8, 2020

https://www.nationalww2museum.org/war/articles/atomic-bombings-75-ian-w-toll

212 Goralski 417 – Aug. 15, V-J Day. Hirohito on the radio for the first time to order Japanese to lay down their arms.

419 – Aug. 25, Radio Tokyo reports cases of people killing themselves in front of the royal palace. Aug. 27, American fleet of 23 aircraft carriers, 12 battleships, 26 cruisers,116 destroyers and escorts, 12 submarines, and 185 smaller ships assembled into Sagami Bay near Tokyo, then to Tokyo Bay.

Toland 959, 974 – Suicides in front of the imperial palace.

944 – Over several pages, Toland tells the story of how the emperor's speech was recorded and smuggled to the radio station for broadcast the next day. Those who opposed surrender hoped to steal the two recordings. Hirohito's voice had been recorded only one before, by accident in 1928, when he was some distance from the microphone that caught him speaking. Few heard the earlier broadcast, but nearly all Japanese citizens stood at attention to listen to the surrender.

Vigeland, Tess. "Hirohito's Speech: The Surrender Of Japan's 'Living God,'" "All Things Considered," Aug. 15, 2015 https://www.npr.org/2015/08/15/432399603/hirohitos-speech-the-surrender-of-japans-living-god

Tenney 191 – All the records of those who were surrendered in 1942 had been destroyed. The U.S. Army interviewed the newly freed men, told them about their promotions and medals, and provided new uniforms. Most also had medical exams at a camp near Manila.

213 Tenney 188, 191 – Returning to Manila for processing, new uniforms.

214]Coleman 87 – He was on burial detail at Camp O'Donnell and described mass graves and the difficulty of keeping track of who was being buried. Japanese guards had taken many POWs' dog tags.

"Background on the recovery of the remains of the men of the *Enoura Maru*," *Enoura Maru* Hellship Memorial Stone Dedication: National Memorial Cemetery of the Pacific, August 15, 2018: In memory of: *Oryoku Maru, Enoura Maru, Brazil Maru*. P 10.

Glor, Jeff. CBS This Morning, "The Search for answers about 'hell ships of World War II,"

Top of Form

May 29, 2021 https://www.youtube.com/watch?v=fxe-igDzxrE

Hirrel, Dr. Leo P. "The beginnings of the Quartermaster Graves Registration Service," U.S. Army, July 8, 2014. History of the American Graves Registration Service.

https://www.army.mil/article/128693/the_beginnings_of_the_quartermaster_graves_registration_service

Holik, Jennifer. "WWII American Graves Registration Service: And the Story of Sergeant John J. Kubinski," World War 2 History Short Stories, May 16, 2019.

Traces the beginning of the Graves Registration Service to the Civil War.

https://www.ww2history.org/war-in-europe/wwii-american-graves-registration-service-and-the-story-of-sergeant-john-j-kubinski/

Tom Kelly, the great-grandson of Ty's aunt, Meta Kokjer Key, and the author of several local histories, has been working with local officials at the Fridhem Cemetery near Funk to erect a memorial to Ty near Hans and Shy's graves.

215 "History," American Battle Monuments Commission. https://www.abmc.gov/about-us/history

Manila American Cemetery, American Battle Monuments Commission. https://www.abmc.gov/Manila

Epilogue

216 The letter to Jim DeWolf did not survive.

217 Dallek 238-239 – Should Roosevelt ease up on oil to delay or prevent war.

218 Robert G. Bjoring obituary, April 26, 1920 to June 8, 2006

https://www.legacy.com/obituaries/sanantonio/obituary.aspx?n=robert-g-bjoring&pid=88866441

219 Michno 316 – The *Arisan Maru* sinking. 1,792 died.

Daws 293 – *Arisan Maru* sunk, eight survived.

220 Scott 174-176 – Japanese guards walked away from Bilibid.

79-81 – In Palawan, POWs murdered.

Knox 338 – When MacArthur reached Luzon in January 1945, he realized all able-bodied POWs were gone. Sick POWs at Bilibid and Cabanatuan liberated in early 1945, plus 5,000 civilians and nurses at Santo Tomas.

221 Scott 133 – Japan did not declare Manila an open city.

422 – Fighting in Manila resulted in the deaths of 16,665 Japanese forces, nearly all whom had remained in Manila. The United States lost 1,010 soldiers, and 5,565 were wounded. An estimated 100,000 Philippines civilians died, at hands of Japan or because of U.S. attacks on Japanese positions.

416-417 – Japanese commander in the city, said at the end that he would commit suicide, and those under his command could do the same or surrender.

222 Groom 415 – 1942 + "Multitudes of people who had never been out of their own states or even counties now traveled hundreds of even thousands of miles to settle into war work. Women from Appalachia found themselves in places like Pittsburgh or Richmond or Indianapolis, working on assembly lines; blacks from southern farmlands migrated to New Orleans, Houston, Mobile, Chicago, and Detroit to build warships, tanks, and airplanes."

"Millions of soldiers and sailors finally got to see the world, and so formed a broader, more enlightened picture for themselves. Many never returned to their home places afterward, thus resulting in a significant shift in the U.S. population."

223 Interview, K.C. Schaack. Her family owned and ran the hotel in 2017 when I visited. They sold it to one of their long-time employees in late 2021.

224 The number of living World War II veterans dropped from 300,000 in 2020 to about 240,000 in 2021. The number in 2022 declined to just over 167,000, with 180 veterans dying daily.

Schaeffer, Katherine, "On 75th anniversary of V-E Day, about 300,000 American WWII veterans are alive," Pew Research Center.

https://www.pewresearch.org/fact-tank/2020/05/08/on-75th-anniversary-of-v-e-day-about-300000-american-wwii-veterans-are-alive/

"WWII Veteran Statistics: The Passing of the WWII Generation"

The National World War II Museum, 2020 – 300,00 remained alive. In 2021, the number was down to 240,329.

https://www.nationalww2museum.org/war/wwii-veteran-statistics

Figures from the Veterans Administration are updated annually, in September.

"The Passing of the World War II Generation," https://www.nationalww2museum.org/war/wwii-veteran-statistics

Michno vii – Post war POWs: "The onus of having been captured played a part in their silence, as did the unwillingness to relive a bad experience. ... Many felt guilty about surviving; others were ashamed of what they had been forced to do in order to survive."

307 – Most POWSs scarred by the experience. Some did better than others after the war. Worse health than the population at large.

Daws 376-379 – Difficult adjustments at home. "It made the POWs strangers among their own folks." Some committed suicide.

358 – one former POW went to a movie about the war, and when a plane flew across the screen, he hid under his seat.

384 – most POWs were still young after the war, but had physical issues for many years. Death rate was far higher than civilians their age, and for war vets who were not POWs.

Martin and Stephenson 287-289 – No banners to welcome them home, "We had been ...marked as cowards." Too many winners to acknowledge losers. About 30 percent of the 27th Bomb Group survived the war. Ninety percent of German POWs survived.

Tenney xv-xv – Felt like loser when arrived home in Oct. '45. His brother took notes of all he said and he had some notes. Decided to write memoir.

Xvii – still has nightmares about the Death March. (The first edition of his book was published in 1995.)

198 - It still seemed as if the war in Europe had been more important and VE day was more important than VJ day. Americans wanted to get on with their lives, even though Pacific POWs were held for another three months. Now, they didn't even want to be seen with their uniforms on. "We did not know

what to do or how to handle it. If we could have crawled into a hole and pulled it in after us, I think we would have."

Interviews with sisters Darcy Pattison and Elleen Hutcheson: Their father's incessant talk about Bataan, and visits from friends who needed to talk about the war – often with plenty of alcohol – was hard on the family. It played a part in their parents' divorce. Later in life, he and Elleen lived near each other and she and her children heard the stories from a more distant perspective.

Henry Thomas Foster, who was born in New Mexico, died in 1983 at age 77.

225 Rosario, Isabella, "Code Switch: The Unlikely Story Behind Japanese Americans' Campaign For Reparations." NPR, March 24, 20206

https://www.npr.org/sections/codeswitch/2020/03/24/820181127/the-unlikely-story-behind-japanese-americans-campaign-for-reparations

Herszenhorn, David "Americans Held Hostage in Iran Win Compensation 36 Years Later," The New York Times, December 24, 2015.

https://www.nytimes.com/2015/12/25/us/politics/americans-held-hostage-in-iran-win-compensation-36-years-later.html

Gearan, Anne, "40 years later, a dwindling band of Iran hostages awaits a promised payment," Washington Post, Sept. 21, 2021.

https://www.washingtonpost.com/politics/2021/09/26/40-years-later-dwindling-band-iran-hostages-awaits-promised-payment/

Reynolds, Gary K., "U.S. Prisoners of War and Civilian American Citizens Captured and Interned by Japan in World War II: The Issue of Compensation by Japan,"

Information Research Specialist, Information Research Division, Congressional Research Service, The Library of Congress, Prepared for Members and Committees of Congress, Updated

https://www.everycrsreport.com/reports/RL30606.html

Joseph, Jennifer, "POWs Left in the Cold: Compensation Eludes American WWII Slave Laborers for Private Japanese Companies Slave Laborers for Private Japanese Companies." Pepperdine Law Review, Review Volume 29 Issue 1 International Law Weekend – West,

https://digitalcommons.pepperdine.edu/cgi/viewcontent.cgi?article=1320&context=plr

Kotler, Mindy email: Congressional Gold Medal campaign

"Technically, the spirit and intent of the Filipino Veterans of WW II Congressional Gold Medal Act of 2015 is for only Filipinos soldiers and guerrillas and their American officers who had active-duty status in the Philippines during WWII (that means the entire war)."

Daws 383 – POWs, expected a grateful country "to shower upon" them all kinds of free things, instead, they got back pay and "some miserable pittance per day of prison camp."

389 – Pay to others: Iranian hostages. Damages to Vietnam protesters wrongly imprisoned. Japanese American wrongly imprisoned at home got more than POWs. No comparison between camps in United States – which were harsh – and cruel prison camps run by Japan.

Michno 281-282 – Many more civilians were killed in WWII than combatants – 55 million civilians, and 22 million combatants.

"Civilian War Dead Roll of Honour 1939 – 1945," Westminster Abbey, London

https://www.westminster-abbey.org/abbey-commemorations/commemorations/civilian-war-dead-roll-of-honour-1939-1945

"Research Starters: Worldwide Deaths in World War II," The National World War II Museum. https://www.nationalww2museum.org/students-teachers/student-resources/research-starters/research-starters-worldwide-deaths-world-war

Department of Defense, "75 Years of the GI Bill: How Transformative It's Been"

Jan. 9, 2019. https://www.defense.gov/Explore/Features/story/Article/1727086/75-years-of-the-gi-bill-how-transformative-its-been/

"The Post War United States, 1945-1968," U.S. History Primary Source Timeline, Library of Congress

http://www.loc.gov/teachers/classroommaterials/presentationsandactivities/presentations/timeline/postwar/

The recovery benefited White vets and civilians more than Black vets and civilians.

Pruitt, Sarah, "The Post World War II Boom: How America Got Into Gear," History,

May 14, 2020

https://www.history.com/news/post-world-war-ii-boom-economy

Bohanon, Cecil, "Economic Recovery: Lessons from the Post-World War II Period," Mercatus Center, George Mason University, Sept. 10, 2012

https://www.mercatus.org/publication/economic-recovery-lessons-post-world-war-ii-period

Fry, Richard, "Millennials overtake Baby Boomers as America's largest generation," Pew Research Center, April 28, 2020 https://www.pewresearch.org/fact-tank/2020/04/28/millennials-overtake-baby-boomers-as-americas-largest-generation/

Michno 307 – Not all promises of help and health care for POWs were kept. They had worse health and died at a younger age than others. Japan never paid compensation or admitted brutality, although Germany did admit its actions.

226 Mead Ordnance plant

Archive Photos Mead Ordnance Plant, Nov. 15, 2012

https://journalstar.com/promo/homepage/archive-photos-nebraska-ordnance-plant-at-mead/collection_85236bdd-9916-59c1-84c2-b0d2fb44beb2.html

McKee, Jim, "Cornhusker ordnance plants helped America win WWII," The Lincoln Journal-Star, Jan 12, 2014 Updated Jan 22, 2015. History of ordnance plants in Nebraska, and origin of WOWs Women Ordnance Workers.

http://journalstar.com/news/local/jim-mckee-cornhusker-ordnance-plants-helped-america-win-wwii/article_a944142c-9c1f-52a3-a016-416476a00741.html

Cohen, Shera, "WOWs (Women Ordnance Workers) to the Rescue!" Feb. 13, 2012, National Park Service. https://www.nps.gov/spar/learn/news/wows-to-the-rescue.htm

Schiller, Joyce K., June 30, 2011, "She's a WOW," Rockwell Center for American Visual Studies https://www.rockwellcenter.org/essays-illustration/shes-a-wow/

"Rosie the Riveter more than a poster girl," Army Military History https://goordnance.army.mil/history/rosie.html

The Associated Press, "NU begins cleanup at former ordnance plant near Mead,"

The Denver Post, September 16, 2007.

Mentions plant's operations in World War II and Korean wars, then tells about cleanup.

https://www.denverpost.com/2007/09/16/nu-begins-cleanup-at-former-ordnance-plant-near-mead/

Hilfik, Genevieve. "Nebraska...Our Towns: Mead--Saunders County," University of Nebraska-Lincoln, undated. https://casde.unl.edu/history/counties/saunders/mead/index.php

"About Midland University," Mission and History.

In 1962, Luther College in Wahoo merged with Midland University in Fremont, Nebraska, about 20 miles north of Wahoo, where there is no longer a campus.

https://www.midlandu.edu/about/mission-history/

227 James G. DeWolf

Nov. 17, 1919 – Aug. 28, 1995

Buried Old Saltillo Cemetery, Texas

https://www.findagrave.com/memorial/26146732/james-gilman-dewolf

He flew the B-24: The Great Liberator

https://www.lockheedmartin.com/en-us/news/features/history/b-24.html

Hiram Messmore

January 31, 1919 - November 7, 1993.

https://www.ancestry.ca/discoveryui-content/view/68555507:60525

https://www.ancestry.ca/discoveryui-content/view/68548042:60525

https://www.findagrave.com/memorial/110691364/hiram-allison-messmore

Beatrice Times July 1945 – promoted to major, about to enter USC law school

"Lt. Hiram Messmore 'Jumped' by 30 Jap Planes But Pulled Thru"

Lincoln Journal, Nov. 10, 1942

19th Bombardment Group History

http://www.historyofwar.org/air/units/USAAF/19th_Bombardment_Group.html

228 Manila American Cemetery, American Battle Monuments Commission,

https://www.abmc.gov/Manila

Cemeteries and Monuments, American Battle Monuments Commission.

"ABMC administers, operates and maintains 26 permanent American military cemeteries and 32 federal memorials, monuments and markers, which are located in 17 foreign countries, the U.S. Commonwealth of the Northern Mariana Islands, and the British Dependency of Gibraltar; four of the memorials are located within the United States."

The site has a search tool to find the grave or memorial to any of the cemeteries and monuments the commission administers.

https://www.abmc.gov/cemeteries-memorials

Alvan Stanley Ose

December 2, 1918 to December 15, 1944. or January 9, 1945.

Memorialized at Manila American Cemetery and the Story City (Iowa) Cemetery

Awarded the Purple Heart and the Bronze Star

There has been some disagreement about when and where he died—either on the *Oryoku Maru* during the bombing raid in Subic Bay or when the *Enoura Maru* was bombed in Formosa. He was awarded the Purple Heart. His body was never found.

http://www.storycityhistory.org/ose-alvan-s.html

https://www.findagrave.com/memorial/98345457/alvan-stanley-ose

https://www.abmc.gov/decedent-search/ose%3Dalvan

George S. Davis

Born 1917, Died Jan. 9, 1945.

First lieutenant, 91st Squadron, 27th Bomb Group.

He was on the *Oryoku Maru* and died when the *Enoura Maru* sank. He is listed as missing in action. His remains may be in the graves of unidentified dead at Punchbowl. He is memorialized at Manila American Cemetery.

Of Cumberland County, Maine, which includes Portland

https://www.findagrave.com/memorial/56772544/george-seiders-davis

http://www.russpickett.com/history/mecumb.htm

229 America Counts Staff, "Nebraska Population Neared 2 Million in 2020," Nebraska 2020 Census, U.S. Census Bureau https://www.census.gov/library/stories/state-by-state/nebraska-population-change-between-census-decade.html

The state's population was 1,961,504 in 2019, up from 1,358,000 in 1940.

Drozd, David and Deichert, Jerry. "Nebraska Historical Populations, Quick Reference Tables,"
Center for Public Affairs Research, University of Nebraska at Omaha (January 2018).

https://www.unomaha.edu/college-of-public-affairs-and-community-service/center-for-public-affairs-research/documents/nebraska-historical-population-report-2018.pdf

Lincoln's population in 1940 was 81,984. In 2019 it was 289,102, with a metro area population in 2018 of 334,920. The Omaha metro area makes up nearly half of the state's population. The city of Omaha population was 223,844 in1940 an grew to 475,862 in 2019. The Omaha metro area population of 967,604 in 2018 includes three counties in Iowa.City of Omaha, "Metro Omaha Population Nears 1,000,000," (August 12, 2021).

The city predicts that number will top one million by 2024.

https://www.cityofomaha.org/latest-news/804-metro-omaha-population-nears-1-000-000

Nebraska State Government, "County Populations 1860 to 1950," April 2002

https://opportunity.nebraska.gov/wp-content/uploads/2020/03/COUNTY-POPULATIONS_1860-1950.pdf

Meanwhile, the population of Grant County, home to Hyannis, was 611 in 2020, down from 1,327 in 1940.

230 Gjerde, Jon, "The Scandinavian Migrants," p. 85. a chapter that is part of *The Cambridge Survey of World Migration*, (1995). That chapter is based on information from the *Harvard Encyclopedia of American Ethnic Groups*, (1980), edited by

Thernstrom, Stephan. The information is part of an exhibition at the National Nordic Museum in Seattle, Washington. https://www.nordicmuseum.org/

Danish Immigration, Museum of Danish America, adapted from the museum's core exhibition "Across Oceans Across Time." https://www.danishmuseum.org/explore/danish-american-culture/immigration

Says that between 1820 and 1990 more than 375,000 Danes came to the United States, a longer timeframe that cited by the Nordic Museum.

Danish American History, National Foundation for Danish America

https://danishamerica.org/history

The foundation's website traces Danish immigration from the first explorers in the seventeenth century. But the greatest wave of immigrants—approximately 300,000—arrived between 1864 and 1930, primarily because of economics.

231 Hansen, Maria and Osterhout, Elva Kokjer, "Kokjer Relatives," unpublished family history. Undated.

232 Object Record Number: 10106-1-(1-2). Dress, two piece, wedding, 1883, green silk, Sarah Kokjer.

https://nebraskahistory.pastperfectonline.com/webobject/81F965AE-F947-43DC-BF3E-914688608667

Although the dress looks green in photo, it is not quite that green when viewed in person. It has faded from the original color, "ashes of rose." Sarah Malina Alice Hartwell Kokjer, went by Malina, not Sarah.

Hans Madsen Kokjer, 1857-1932, letters. [RG1141.AM]. Correspondence: 1874-1918, 1973-1979, in Danish with an English translation by a Danish cousin. The early letters are between Hans and a cousin. The later set are letters between Hans' youngest son, Emerson, and a Danish cousin.

https://history.nebraska.gov/collections/hans-madsen-kokjer-1857-1932-rg1141am

Clarks history book. *Heritage of Clarks, Nebraska*, Compiled and Edited by Clarks Bicentennial Heritage Committee, 1865-1976.

Thomas E. Kelly papers.

Index

Page numbers in italics refer to photographs. N. or nn refer to end notes.

N

R

Photo by K'Von A. Jackson

Jody Beck grew up in Lincoln, Nebraska, the oldest of five children born to Leo J. Beck Jr. and Phyllis Kokjer Beck. Ty's parents were occasional visitors to the Beck home, but both died when Beck was in college and before she knew about Ty's story.

She was a reporter for The Washington Star, an assignment editor at WRC-TV, taught journalism at the University of Maryland-College Park, and was director of the Scripps Howard Foundation's Semester in Washington Program.

She earned a bachelor's degree from the University of Nebraska-Lincoln, and a master's degree from the University of Maryland, both in journalism. She lives in Washington, D.C.

Contact her, read blog posts, and leave questions or reviews at jody-beck.com. Follow her on X (Twitter) @JodyBeckDC, or on Threads jodybeck26.

www.ingramcontent.com/pod-product-compliance
Lightning Source LLC
Chambersburg PA
CBHW071135130626
46553CB00004B/1390